Intratheater Airlift Functional Needs Analysis (FNA)

John Stillion, David T. Orletsky, Anthony D. Rosello

Prepared for the United States Air Force
Approved for public release; distribution unlimited

PROJECT AIR FORCE

The research described in this report was sponsored by the United States Air Force under Contract FA7014-06-C-0001. Further information may be obtained from the Strategic Planning Division, Directorate of Plans, Hq USAF.

Library of Congress Cataloging-in-Publication Data

Stillion, John.
Intratheater airlift functional needs analysis (FNA) / John Stillion, David T. Orletsky, Anthony D. Rosello.
p. cm.
Includes bibliographical references.
ISBN 978-0-8330-4755-7 (pbk. : alk. paper)
1. Airlift, Military—United States—Planning. 2. Transportation, Military—United States. 3. United States. Air Force—Transportation. I. Orletsky, David T., 1963– II. Rosello, Anthony D. III. Title.

UC333.S75 2011
358.4'4—dc22

2009028672

Published 2011 by the RAND Corporation
1776 Main Street, P.O. Box 2138, Santa Monica, CA 90407-2138
1200 South Hayes Street, Arlington, VA 22202-5050
4570 Fifth Avenue, Suite 600, Pittsburgh, PA 15213-2665
RAND URL: http://www.rand.org/
To order RAND documents or to obtain additional information, contact
Distribution Services: Telephone: (310) 451-7002;
Fax: (310) 451-6915; Email: order@rand.org

Preface

This functional needs analysis (FNA) for U.S. Air Force (USAF) intratheater airlift is the second in a series of three documents that together comprise a capabilities-based assessment (CBA) required as part of the Joint Capabilities Integration and Development System (JCIDS). The first document in the series, the functional area analysis (FAA) identified the operational tasks, conditions, and standards needed to achieve military objectives—in this case, certain intratheater airlift missions.[1] The second, this FNA, assesses the ability of the current assets to deliver the capabilities identified in the FAA. The third and final document in this series, the functional solution analysis (FSA), is forthcoming.[2] The FSA will provide an operationally based assessment of approaches to address the capability gaps identified in the FNA.

The research described in this monograph was sponsored by Maj Gen Thomas P. Kane, Director, Plans and Programs, Headquarters, Air Mobility Command, Scott Air Force Base, Illinois. The work was performed within the Force Modernization and Employment program of RAND Project AIR FORCE as part of a fiscal year 2006 study, "Improving Air-Ground Integration, Interoperability, and Interdependence."

---

[1] David T. Orletsky, Anthony D. Rosello, and John Stillion, *Intratheater Airlift Functional Area Analysis (FAA)*, Santa Monica, Calif.: RAND Corporation, MG-685-AF, 2011.

[2] David T. Orletsky, Daniel M. Norton, Anthony D. Rosello, William Stanley, Michael Kennedy, Michael Boito, Brian G. Chow, and Yool Kim, *Intratheater Airlift Functional Solution Analysis (FSA)*, Santa Monica, Calif.: RAND Corporation, MG-818-AF, 2011.

RAND Project AIR FORCE

RAND Project AIR FORCE, a division of the RAND Corporation, is the U.S. Air Force's federally funded research and development center for studies and analyses. PAF provides the Air Force with independent analyses of policy alternatives affecting the development, employment, combat readiness, and support of current and future aerospace forces. Research is conducted in four programs: Force Modernization and Employment; Manpower, Personnel, and Training; Resource Management; and Strategy and Doctrine.

Additional information about PAF is available on our website: http://www.rand.org/paf/

Contents

Figures

Tables

Summary

This document presents the results of the FNA that RAND Project AIR FORCE produced for USAF intratheater airlift. The FNA is the second in a series of three analyses that together comprise a CBA required as part of the JCIDS. The first, the FAA, identified the operational tasks, conditions, and standards needed to achieve military objectives—in this case, certain intratheater airlift missions.[1]

The CBA itself was initiated to analyze a potential deficiency in intratheater airlift capability. It was prompted by concerns that demands from the ongoing global war on terrorism and new U.S. Army operational concepts under consideration might result in a shortfall in the USAF's capabilities to deliver personnel and equipment to increasingly numerous and dispersed theater operating locations. Routine supply of a moderately sized ground combat force using the existing intratheater airlift system is challenging.[2]

The FAA identified three broad operational mission areas for intratheater airlift.[3] These mission areas are the ability to provide (1) routine sustainment; (2) time-sensitive, mission-critical (TS/MC) resupply; and (3) maneuver capabilities to U.S. and allied forces across

---

[1] Orletsky, Rosello, and Stillion, 2011.

[2] While the U.S. Army has a limited fixed-wing airlift capability (currently consisting primarily of C-23 and C-12 aircraft), the USAF has primary responsibility for joint air mobility missions and the bulk of the joint capability for fixed-wing air mobility.

[3] Meeting at Air Mobility Command, December 8, 2005, and subsequent discussions with USAF and U.S. Army personnel.

all operating environments. The FNA assesses the ability of current assets to deliver the capabilities that the FAA identified.

Routine sustainment is defined as the steady-state delivery of required supplies and personnel to units. TS/MC resupply is defined as the delivery of supplies and personnel on short notice, outside the steady-state demands. The maneuver mission is defined as the transport of combat teams around the battlefield using the intratheater airlift system. These three operational mission areas have different characteristics and impose different requirements on the intratheater airlift system.

Aggregate Intratheater Airlift Needs and Current Capabilities

Over the past several years, two relevant studies have addressed the total amount of fixed-wing intratheater airlift capability the USAF will need in the future.[4] The largest and most widely known of these studies is the Mobility Capabilities Study (MCS) conducted as a joint effort by the Office of the Secretary of Defense (OSD) Director of Program Analysis and Evaluation (PA&E) and the Joint Staff.[5] As a follow-up to that study, the Intratheater Lift Capability Study (ITLCS) took a more-detailed look at intratheater lift requirements in wartime.[6] Both studies used OSD-approved planning scenarios as the basis for the mobility demand. These scenarios were set in 2012 and used the weapons and force structures planned to be operational at that time.

Although the studies produced ranges for total fixed-wing intratheater airlift needs that were based on differing assumptions about the location, concepts of operation, duration, intensity, and simultaneity of

---

4 A third study is sometimes discussed, but it was based on a planning year of 2005 and was not considered here.

5 U.S. Department of Defense and the Joint Chiefs of Staff, *Mobility Capabilities Study*, Washington, D.C., December 2005, Not Available to the General Public.

6 ITLCS analyzes intratheater airlift in somewhat more detail than the models used in the MCS. The overall intratheater airlift demand identified in ITLCS is only marginally different from the MCS.

the contingencies and conflicts that produce such needs, both found a consistent level of demand. The minimum C-130 total aircraft inventory was determined to be 395 aircraft.

Three models of the C-130 have been assigned to the mobility air forces (MAF): C-130E, C-130H, and C-130J. On January 3, 2007, there were 405 MAF C-130E/Hs on a total aircraft inventory basis. In addition, there were 37 MAF C-130Js, for a total fleet of 442 aircraft.[7] A large and growing portion of the C-130 fleet is either restricted or grounded due to fatigue-related cracking of key structural components of the center wing box. Many older aircraft are currently restricted from carrying useful cargo loads or are grounded because of fears that fatigue-related cracks in their center wing box structures could propagate in flight and result in a catastrophic collapse of the center wing structure. As of January 2007, 45 C-130Es and C-130Hs were operating under flight restrictions because of the risks accumulated fatigue damage pose.[8] These flight restrictions on weight and flight profile are severe. As a result, restricted aircraft have limited use as airlifters. (See pp. 10–14.)

Possible Sources of Increased Intratheater Airlift Demand

Recent intratheater lift studies have established the basic demand, but several other factors could increase the capability needed. These include the desire to minimize vulnerable ground movements in counterinsurgency environments, the dispersed nature of the global war on terror, and emerging Army concepts of operation that stress operational maneuver and resupply by air.

[7] The 442 MAF aircraft do not include the LC-130s and the WC-130s, since these are special-mission aircraft that are specially configured and fly specific nonmobility missions. These aircraft can and do fly Air Mobility Command missions but may not always be available. Further, their special equipment may limit the amount and type of cargo they can carry. As a result, the discussions in this monograph do not include these among the MAF aircraft.

[8] An inspection and repair process can be undertaken to remove these restrictions and return the aircraft to unrestricted operations for several additional years. Not all aircraft pass the inspection or can be repaired economically.

To explore how these new elements could affect the demand for USAF intratheater airlift capacity, we examined two plausible vignettes. We chose these to examine the consequences of operating from a high, hot desert environment and a low jungle environment, using Afghanistan and Indonesia as notional deployment locations. We looked at both routine sustainment and TS/MC resupply of a sizable ground force. Each vignette involved supplying and sustaining a ground force of approximately six brigade combat teams (BCTs). To examine a spectrum of demands, we varied the types of brigades involved.

Routine Sustainment

Although the current Army concept for future operations does not involve large multi-BCT forces operating without a ground line of communication, the trend is toward more-dispersed ground-force operations. Future ground forces will rely on increased aerial distribution.[9] Such concepts as mounted vertical maneuver will likely rely more heavily on aerial sustainment than do current operations. To better understand the possible implications of routine sustainment, we looked at a case that provided 100 percent by air. Our analysis found that routine sustainment of a moderately sized ground combat force using the existing intratheater airlift system is extremely challenging.

In most of the cases we analyzed, the number of C-130s required to supply six BCTs by air represents a very large fraction of the existing C-130 fleet. The CH-47D helicopter fleet faces even greater challenges.

Time-Sensitive, Mission-Critical Resupply

We examined how intratheater airlift assets might be used to ensure that each combat battalion could take delivery on TS/MC resupply items in 8 hours or less. Our analysis shows that the existing intratheater airlift assets can be combined to provide a robust, responsive TS/MC resupply system with a reasonably small commitment of resources. (See pp. 56–57.)

---

[9] See U.S. Army Aviation Center, Futures Development Division, Directorate of Combat Developments, *Army Fixed Wing Aviation Functional Needs Analysis Report*, Fort Rucker, Ala., June 23, 2003b, p. 16-17.

Our analysis also suggests that returns rapidly diminish if resources are allocated to this mission beyond the levels we chose: Additional investments do not further reduce in-transit time. This, combined with the fact that in-transit time accounts for only part of the total time between request and delivery, suggests that it may be more fruitful to invest in improved logistics management processes and procedures to realize substantial reductions in the total TS/MC resupply performance. (See pp. 57–60.)

Further, we determined that, in large countries with sparse airfield infrastructure, significant areas are simply beyond the reach of the existing CH-47 fleet to supply Army maneuver units. (See pp. 47–48.)

Findings

The current requirement for intratheater airlift is 395 C-130s. Current USAF policies are restricting and grounding aircraft because of structural fatigue issues associated with the center wing box. If these policies remain in place and if nothing else is done, the number of unrestricted C-130s available to the USAF is projected to fall below the minimum threshold of 395 early in the next decade. This situation alone is sufficient justification for examining options for increasing USAF intratheater airlift capacity through an FSA. (See pp. 13–14.)

In addition to this identified requirement, several factors could increase the amount of intratheater airlift needed. If routine resupply of a multi-BCT Army unit for an extended period is adopted as an intratheater airlift task, additional airlift assets would be required. Robust TS/MC resupply of a sizable ground force, on the other hand, can be accomplished with relatively few airlift assets. Since routine resupply is not a requirement and since TS/MC resupply takes relatively few assets, the FSA should focus on ensuring that the intratheater airlift fleet continues to meet the 395 C-130 requirement identified in the MCS. This requirement needs to be met in light of the large number of aircraft that are expected to undergo flight restrictions and groundings during the next two decades.

Abbreviations

BCT	brigade combat team
CASCOM	U.S. Army Combined Arms Support Command
CBA	capabilities-based assessment
CJCSI	Chairman of the Joint Chiefs of Staff Instruction
CONOPS	concept of operations
CWB	center wing box
EBH	equivalent baseline hour
FAA	functional area analysis
FNA	functional needs analysis
FOL	forward operating location
FSA	functional solution analysis
ITLCS	Intratheater Lift Capability Study
JCIDS	Joint Capabilities Integration and Development System
KIAS	knots indicated airspeed

L1	landing surface category one: a major air base with long, hard-surface runways and significant aircraft parking and servicing infrastructure outside the country in which military operations are being conducted
L2	landing surface category two: an airfield with relatively long and wide, hard-surface runways inside the country in which military operations are ongoing
L3	landing surface category three: airfields are generally shorter, narrower and are in many cases unpaved austere landing surfaces
LOC	line of communication
MAF	mobility air forces
MCS	Mobility Capabilities Study
MVM	mounted vertical maneuver
OSD	Office of the Secretary of Defense
PA&E	Office of the Secretary of Defense/Program Analysis and Evaluation
PFPS	Portable Flight Planning Software
PMAI	primary mission aircraft inventory
STAR	scheduled theater airlift
TAI	total aircraft inventory
TRADOC	Training and Doctrine Command
TS/MC	time-sensitive, mission-critical
UA	units of action
USAF	U.S. Air Force

CHAPTER ONE

Introduction, Purpose, and Scope

This functional needs analysis (FNA) is the second in a series on U.S. Air Force (USAF) intratheater airlift that together comprise a capabilities-based assessment (CBA) required as part of the Joint Capabilities Integration and Development System (JCIDS).[1] The preceding functional area analysis (FAA) identified the operational tasks, conditions, and standards needed to achieve military objectives—in this case, certain intratheater airlift missions.[2] According to Chairman of the Joint Chiefs of Staff Instruction (CJCSI) 3170.01E, the FNA

> assesses the ability of the current and programmed warfighting systems to deliver the capabilities the FAA identified under the full range of operating conditions and to the designated measures of effectiveness.[3]

---

[1] Department of Defense Instruction (DoDI) 5000.2, *Operation of the Defense Acquisition System*, May 12, 2003, was the baseline for our understanding of JCIDS and CBAs. Note that the Department of Defense (DoD) updated this instruction in 2008, well after we completed the groundwork for our analysis.

[2] David T. Orletsky, Anthony D. Rosello, and John Stillion, *Intratheater Airlift Functional Area Analysis (FAA)*, Santa Monica, Calif.: RAND Corporation, MG-685-AF, 2011.

[3] CJCSI 3170.01E, *Joint Capabilities Integration and Development System*, May 11, 2005, p. A-4. Since we began our work, this instruction has been revised twice. Much of the material describing the CBA process and the F-series documents has been split off into a second document: Chairman of the Joint Chiefs of Staff Manual (CJCSM) 3170.01C, *Operation of the Joint Capabilities Integration and Development System*, May 1, 2007.

The third and final document in this series, the functional solution analysis (FSA), is forthcoming.[4] That document will provide an operationally based assessment of approaches to address the capability gaps identified in the FNA.

The broad objective of the "F-series" documents (the FAA, FNA, and FSA) is to determine whether a materiel solution is required to address specific shortfalls in military capabilities or whether modifying other aspects of the system could resolve the shortfall.

CJCSI 3170.01E establishes the policies and procedures of JCIDS as specified in U.S. Code.[5] JCIDS and its validated and approved documentation provide the chairman advice and assessments in support of the laws governing military acquisition. CJCSI 3170.01E also provides joint policy, guidance, and procedures for recommending changes to existing joint resources when these changes are not associated with a new defense acquisition.

CJCSI 3170.01E provides a top-down process for identifying needed capabilities that begins with high-level guidance from the National Security Strategy and the National Defense Strategy.[6] Individual service concepts of operations (CONOPS) and the Family of Joint Future Concepts, both developed from the national strategies, also inform and may initiate a need for new capabilities.[7]

JCIDS vets alternative approaches to closing potential capability gaps through a standardized analysis process. The results of the analysis process are then used to make recommendations on how to best acquire the needed capability through possible changes in doc-

---

[4] David T. Orletsky, Daniel M. Norton, Anthony D. Rosello, William Stanley, Michael Kennedy, Michael Boito, Brian G. Chow, and Yool Kim, *Intratheater Airlift Functional Solution Analysis (FSA)*, Santa Monica, Calif.: RAND Corporation, MG-818-AF, 2011.

[5] CJCSI 3170.01E, 2005.

[6] George W. Bush, *The National Security Strategy of the United States of America*, Washington, D.C.: The White House, September 2002, and U.S. Department of Defense, *The National Defense Strategy of the United States of America*, Washington, D.C., June 2008.

[7] Joint Chiefs of Staff, Joint Experimentation, Transformation, and Concepts Division (JS/J7), *JOpsC Family of Joint Concepts—Executive Summaries*, briefing, August 23, 2005.

trine, organization, training, materiel, leadership and education, personnel, facilities, and policy.

This CBA was initiated to analyze a potential shortfall in intratheater airlift capability. It was prompted by concerns that the demands of the ongoing global war on terrorism and the new operational concepts the U.S. Army is considering might create a shortfall in USAF capabilities to deliver personnel and equipment to increasingly numerous and dispersed theater operating locations.

Although large forces consisting of multiple brigade combat teams (BCTs) operating without a ground line of communication (LOC) are not the current Army concept for future operations, the trend is toward more-dispersed operations of ground forces. Future ground forces will rely on increased aerial distribution.[8] Such concepts as mounted vertical maneuver (MVM) may rely more heavily on aerial sustainment than current operations. The Army fixed-wing aviation FNA states that future ground forces will rely increasingly on aerial distribution.[9] It further discusses the dispersed nature of the future operational concept, in that ground forces will "be inserted into an OA [Operational Area] and be supported with some portion (and in surges, possibly all) of their supplies by air." Such concepts as MVM that may rely more heavily on aerial sustainment are being discussed. Any significant shortfall or even delay in supply is likely to reduce ground combat power, thus postponing or canceling ground operations. Efficient delivery of cargo and personnel to numerous, dispersed locations using USAF fixed-wing aircraft is often discussed as a potential solution. This analysis attempts to span the spectrum of potential demands and to investigate the implications for intratheater airlift across a wide range.

Traditional airlift of personnel and materiel using airdrop or airland methods has been employed on numerous occasions during recent operations when surface LOCs were either not available, inadequate, or threatened by adversary action or when cargo needed to arrive quickly.

---

[8] See U.S. Army Aviation Center, Futures Development Division, Directorate of Combat Developments, *Army Fixed Wing Aviation Functional Needs Analysis Report*, Fort Rucker, Ala., June 23, 2003b, p. 16-17.

[9] See U.S. Army Aviation Center, 2003b, p. 16-17.

Effective support of current and future ground combat operations may require capabilities that do not exist in the current programmed USAF airlift fleet. For example, capabilities for operation from short and rough fields or for aircraft survivability required to support future Army CONOPS could be well beyond the capabilities of the current USAF intratheater airlift fleet.

The FAA identified three broad operational mission areas for the intratheater airlift:[10] the abilities to provide (1) routine sustainment; (2) time-sensitive, mission-critical (TS/MC) resupply; and (3) maneuver capabilities to U.S. and allied forces across all operating environments.

Routine sustainment is defined as the steady-state delivery of required supplies and personnel to units. TS/MC resupply is defined as the delivery of supplies and personnel on short notice, outside the steady-state demands. The maneuver mission is defined as the transport of combat teams around the battlefield using the intratheater airlift system. These three operational mission areas have different characteristics and impose different requirements on the intratheater airlift system.

Chapter Two summarizes the capabilities and tasks identified in the FAA as required of the Global Mobility System within the framework of national, joint, USAF, and Army operational concepts. Chapter Three discusses the amount of intratheater lift capability the USAF will need based on the results of the Mobility Capabilities Study (MCS),[11] and compares this capability need to the projected capability of the USAF C-130 fleet. Chapter Four describes an analytical methodology and presents representative operational vignettes to illustrate how the intratheater airlift capabilities described in Chapters Two and Three might be used in the future. Chapter Five describes the results of the analysis, while the final chapter presents our overall conclusions.

---

[10] See Orletsky, Rosello, and Stillion, 2011, for a more-detailed description of the methods and sources used to derive these three mission areas.

[11] U.S. Department of Defense and the Joint Chiefs of Staff, *Mobility Capabilities Study*, Washington, D.C., December 2005, Not Available to the General Public.

CHAPTER TWO

Platform Capability Needs and Conditions

Introduction

The FAA used multiple sources for input and guidance.[1] The FAA provides a complete discussion of the documents, which included the National Security Strategy, the National Defense Strategy, the Joint Operations Concepts Family, the U.S. Air Force Transformation Flight Plan 2004, the Global Mobility CONOPS, and Army Vision. We also consulted the Universal Joint Task List and the Air Force Master Capabilities Library.

Intratheater Airlift Tasks

For the FAA, we used the guidance documents, the Universal Joint Task List, and the Master Capabilities Library to select the tasks, conditions, and standards that are important for intratheater airlift missions. Both routine sustainment and TS/MC resupply have the following relevant tasks:

- transport supplies and equipment to points of need
- conduct retrograde transport of supplies and equipment
- transport replacement and augmentation personnel
- evacuate casualties.

[1] Orletsky, Rosello, and Stillion, 2011, pp. 5–23 and pp. 32–34.

Conditions

Although the guidance documents do not specify a set of conditions under which these tasks must be accomplished, attributes and conditions are discussed throughout the guidance documents. Some of these attributes and/or conditions occur in multiple guidance documents. We identified the following conditions that the Air Force identified as important and should be considered in this CBA:

- adverse weather (e.g., low visibility, temperature extremes, etc.)
- multiple, simultaneous, distributed decentralized battles and campaigns
- antiaccess environment
- support forces operating in and from austere or unimproved locations
- degraded environments (weapons of mass destruction or effect; chemical, biological, radiological, nuclear, and explosive weapons; natural disasters)
- multinational environment[2]
- absence of pre-existing arrangement.

The following attributes and conditions are discussed as positive in the guidance documents:

- establishing the smallest logistical footprint
- delivering with speed, accuracy, and efficiency
- distributing to the point of requirement
- basing flexibly to permit operation across strategic and operational distances.

The guidance documents also specify standards that should be used to evaluate potential gaps in capabilities. The tasks identified

[2] The *multinational environment* includes both support to non–U.S. forces in situations in which U.S. ground forces are not engaged at all or have limited involvement as advisers and support of friendly and allied forces with and without the participation of U.S. ground forces.

above should be accomplished with the following standard capabilities in mind:

- meeting demands for force and materiel movement
- moving forces and materiel throughout a theater optimally
- providing materiel support for current and planned operations.

CHAPTER THREE

Aggregate Intratheater Airlift Needs and Current Capabilities

Over the past several years, three major studies have addressed the total amount of fixed-wing intratheater airlift capability the USAF will need in the future. The largest and most widely known of these was the MCS, which was a joint effort of the Office of the Secretary of Defense (OSD) Director of Program Analysis and Evaluation (PA&E) and the Joint Staff.[1] This broad study addressed strategic airlift, aerial refueling, sealift, and prepositioning capability needs. In addition, the MCS conducted a broad analysis of intratheater transportation requirements, including intratheater airlift.

A follow-up to the MCS, the Intratheater Lift Capability Study (ITLCS), took a more-detailed look at intratheater lift requirements in wartime. Both the MCS and ITLCS used OSD-approved planning scenarios as the basis for the mobility demand. These scenarios are set in 2012 and use weapons and force structures that are planned to be operational at that time.

The objective of the third study, conducted by the Institute for Defense Analyses, was to determine the number of C-130s required to support military missions. With a 2005 time frame, the study is based on the same wartime assumptions as Mobility Requirements Study 2005. Since the MCS has superseded Mobility Requirements Study

[1] The MCS was released on December 19, 2005, and was conducted by PA&E and the Chairman of the Joint Chiefs of Staff in collaboration with the OSD, the Joint Staff, the services, and the combatant commands.

2005, the results of the latter are no longer current and were not used in our analyses.

Current U.S. Air Force Intratheater Lift Capabilities and Limitations

While they used different models and assumptions, the analytic results of both the MCS and ITLCS indicated that the minimum total aircraft inventory (TAI) of USAF C-130s for supplying the necessary intratheater airlift is 395.

Current and Projected USAF C-130 Intratheater Airlift Fleet

Three models of the C-130 are assigned to the mobility air forces (MAF). The C-130Es are the oldest and were produced between 1962 and 1974. Three different versions of the C-130H (the H1, H2, and H3) were produced between 1974 and 1996. The newest C-130 model is the C-130J, which began production in 1996 and is still being built. On January 3, 2007, MAF's TAI included 405 C-130E/Hs and 37 MAF C-130Js, for a total fleet of 442 aircraft.[2] This total includes all Air Force aircraft assigned to active, reserve, and National Guard units whose primary mission is either airlift or training airlift crews. At first glance, this number seems to give a considerable margin of capability beyond the minimum required to meet all intratheater airlift needs. However, a large and growing proportion of the C-130 fleet has either been restricted or grounded because of fatigue-related cracking of key structural components of the center wing box (CWB).

Center Wing Box Restrictions and Groundings

Many of these older aircraft are currently restricted from carrying useful cargo loads or have been grounded because of fears that fatigue-

[2] The 442 MAF aircraft do not include the LC-130s and the WC-130s because these are special mission aircraft that are specially configured and fly specific nonmobility missions. Although these aircraft can and do fly Air Mobility Command missions, they may not always be available. Further, the special equipment may limit the amount and type of cargo that can be carried. As a result, we do not include these among the MAF aircraft throughout this analysis.

related cracks in their CWB structures could propagate in flight and cause catastrophic collapse of the center wing structure. As of January 2007, 45 C-130Es and C-130Hs were operating under flight restrictions because of accumulated fatigue damage. The specific restrictions include the following:

1. The maximum gross operating weight is 139,000 lbs (unrestricted is 155,000 lbs).
2. The maximum zero fuel weight of 90,000 lbs (unrestricted is 130,000 lbs).
3. The minimum landing fuel weight of 15,000 lbs (unrestricted is 4,000 lbs).
4. The maximum airspeed of 190 kts indicated airspeed (KIAS) at or below 2,000 ft above ground level (unrestricted is 318 KIAS).
5. No high-speed, low-level operations are permitted.
6. The maximum maneuver load factor is +2.0g clean and +1.5g with flaps extended (unrestricted is +3.0g clean).
7. Control wheel throw must be limited to ±90 degrees at speeds greater than 185 KIAS.
8. Only primary fuel management is to be used.
9. Flight in moderate or greater turbulence is to be avoided.
10. Abrupt maneuvers are to be avoided.[3]

The key restriction for our purposes is the maximum zero fuel weight of 90,000 lbs. The typical C-130 operating empty weight is between 86,000 and 88,000 lbs. That leaves a margin of only 2,000 to 4,000 lbs for cargo and passengers. This is only 5 to 10 percent of the normal maximum payload of a C-130. So, while restricted aircraft may be used for limited training purposes, they are essentially useless for intratheater air delivery missions.[4]

---

[3] Restrictions listed in Warner Robins Air Logistics Center, ALC/LB, "C-130 Center Wing Status," briefing, February 9, 2005.

[4] While it may seem possible to use the restricted aircraft to perform the TS/MC resupply mission, there are two obstacles to using them in this way. The first is that, even on these "low-payload missions," the payload could exceed 4,000 lbs. Most restricted aircraft have a payload capacity less than 4,000 lbs, with some as low as 2,000 lbs. The second obstacle is

Subtracting the 45 restricted aircraft from the 442 TAI leaves just 397 unrestricted C-130s remaining in the total USAF inventory. If current USAF policies (described below) for restricting and grounding aircraft remain in place and if nothing else is done, the number of unrestricted C-130s available is projected to fall well below the minimum threshold of 395.

Equivalent Baseline Hours. One of the challenges of managing the C-130 fleet is that USAF C-130s fly a diverse suite of mission profiles that put varying levels of stress on the airframe. For example, a simple cargo-delivery mission, in which the aircraft flies at high altitude to its destination and lands, puts less stress on the airframe than a low-altitude mission, which in turn might be less stressful than a number of practice assault landings.

To simplify fleet management and maintenance, USAF and industry developed a method of accounting for the differing levels of wear and tear that result from varying mission profiles. The goal of this method is to convert the number of hours flown on a given mission profile into equivalent baseline hours (EBH) by estimating a severity factor for each mission profile and then multiplying the actual hours flown by this factor. For the high-altitude delivery mission described above, the multiplier would be about one. For a low-altitude mission at a high gross weight and with an assault landing, the multiplier might be 4 or 5. EBH methodology uses detailed records of aircraft weights and flight profiles to establish a multiplier for each mission and then keeps track of the total EBH for each aircraft. The value of the multipliers is derived from fatigue test data and is updated when fatigue cracks are detected during inspections of C-130 airframes.

Aircraft with fewer than 20,000 EBH have very few, if any, fatigue cracks in their CWBs. Any cracks that do exist are typically very small and below the threshold of detection. As EBH levels increase beyond 20,000 hours, single cracks begin to be detected at well-known

---

that these missions are likely to require high-stress maneuvers, such as minimum landing rolls, maximum performance takeoffs, and even flight through moderate or greater turbulence. In short, even for very low-payload missions, the restrictions place severe limits on the operational utility of restricted aircraft.

fatigue-critical locations. Cracks on aircraft with more than 30,000 EBH become more widespread and are not confined to fatigue-critical locations.[5] Current USAF policy is to impose the restrictions outlined earlier on C-130 airframes that reach 38,000 EBH. At 45,000 EBH, aircraft are no longer considered airworthy and are grounded.[6]

Projected Consequences of Groundings and Restrictions on Equivalent Baseline Hours. Figure 3.1 shows how the number of C-130 airframes available to the USAF for intratheater air delivery missions is projected to decline over the next two decades. The figure shows that the projected number of C-130s will fall below the MCS requirement of 395 in 2013. In the analysis, each aircraft can contribute to the requirement until it reaches 45,000 EBH, at which point it is retired.

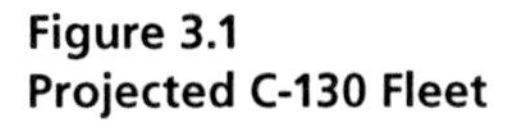

Figure 3.1
Projected C-130 Fleet

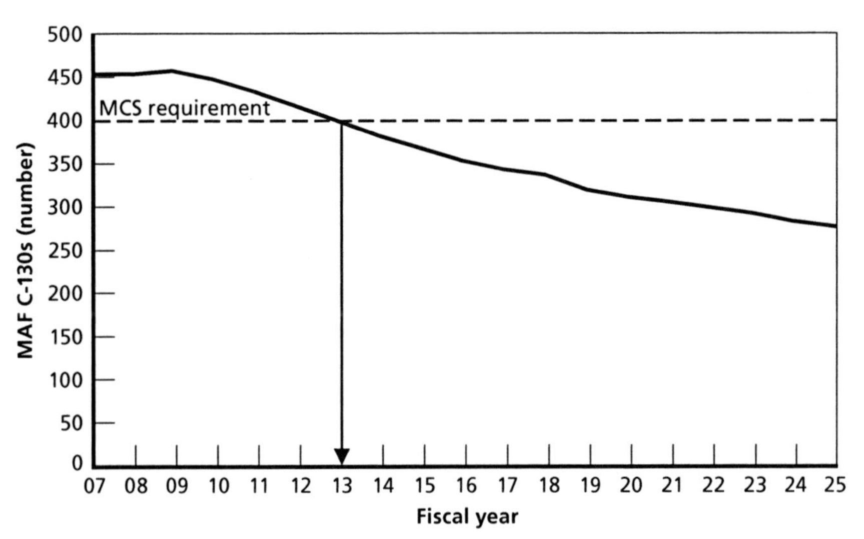

NOTE: This TAI assumes all aircraft undergo Time Compliance Technical Order 1908 and are able to fly to 45,000 EBH.

RAND *MG822-3.1*

---

[5] For a detailed discussion of EBH and C-130 fatigue cracking, see Orletsky et al., 2011, Chapter Two and Appendix A.

[6] Restrictions listed in Warner Robins Air Logistics Center, 2005.

We assumed that all aircraft undergo and successfully complete an inspection and repair procedure at 38,000 EBH allowing them to fly to 45,000 EBH unrestricted. Further, we included the number of new C-130Js that are currently budgeted. These assumptions are fairly optimistic. If some aircraft cannot be repaired after undergoing inspection, the shortfall could occur prior to 2013.[7]

Implications

Given the current policies of flight restrictions and groundings, the number of C-130s will soon fall below the 395 required. It is clear that USAF must now begin investigating the full spectrum of options for maintaining its intratheater airlift capability. This process is begun in the FSA. In the FSA, we will examine in greater detail the EBH methodology, how the current restriction and grounding thresholds were determined, aircraft crew ratios, increased use of simulators, and materiel solutions (including CWB repair and aircraft replacement). We will also examine other doctrine, organization, training, materiel, leadership and education, personnel, and facilities alternatives.[8]

However, before concluding the analysis at hand, the FNA, it is worth examining some of the factors that might cause the demand for intratheater airlift capability to increase beyond the minimum level of 395 C-130s. The next chapter sets out several factors that could increase this demand and examines, in detail, how one factor in particular, proposed U.S. Army CONOPS, could dramatically increase the need for intratheater airlift capability.

---

[7] See Orletsky et al., 2011.

[8] Orletsky et al., 2011.

CHAPTER FOUR

Factors That Could Increase Intratheater Airlift Demand

Introduction

Emerging Needs

In the last chapter, we showed that there was a capability gap between the 395 C-130s needed and the number projected to be remaining in the C-130 fleet over the next 15 years given aircraft retirements. Three related emerging needs not included in the current set of requirements could exacerbate this capability gap by increasing the number of aircraft needed in the future:

- larger percentage of resupply via air
- increasingly dispersed nature of operations
- future Army CONOPS, such as MVM.

This chapter explores these potential emerging demands and presents our analysis of the ability of the current intratheater airlift fleet to meet these increased needs.

Ongoing operations in Afghanistan and Iraq have underscored the vulnerability of ground convoys to attack from irregular and insurgent forces. Especially evident is the vulnerability of supply convoys to ambushes and improvised explosive devices. One of the ways ground supply convoys can be minimized, or in some cases eliminated, is to deliver both routine sustainment and TS/MC resupply items as close as possible to the end user. In addition to eliminating the risks associated with ground resupply operations, aerial resupply offers the potential to reduce delivery times. However, the MCS, ITLCS, and other studies assumed that only about 5 percent of sustainment supplies are delivered to combat units by air.[1] There are no approved plans to replace a significant portion of ground resupply with air delivery. Implementing more aerial resupply could dramatically increase the need for intratheater airlift.

A second possible source of increased intratheater airlift demand is the dispersed nature of the global war on terror, which translates to multiple, simultaneous, decentralized operations scattered across huge areas.

A third source of possible additional intratheater airlift demand is future Army CONOPS that rely heavily on aerial maneuver and resupply of dispersed combat units deep inside enemy-held territory. This is, of course, a "what if" discussion because such future operational concepts as MVM are currently still matters of development and debate. Our focus here is to show the implications of increased reliance on aerial delivery for the intratheater airlift force.

We used two vignettes and two missions to analyze how these potential emerging needs affect the amount of intratheater airlift required. The two vignettes (Afghanistan and Indonesia) were chosen to represent different operational circumstances, including resupply distance within the theater and environmental conditions. The missions we analyzed included resupply of 100 percent of the routine sustainment needs by air and a TS/MC delivery requirement every eight hours to each of 18 locations. In this analysis, we used spreadsheet models to determine the number of aircraft in each of these cases. The rest of this chapter describes our analytical methodology.

---

[1] In Operation Iraqi Freedom and for large Operation Enduring Freedom units, there has never been an attempt to conduct routine sustainment via air. Even at its maximum, air resupply took only a few percent of the trucks off the road (actually, less than 5 percent). The vast bulk of routine sustainment has remained by ground—virtually all food, water, bulk fuel, construction materiel, and ammunition to storage points. RAND colleague Eric Peltz supplied this clarification.

Analysis of Emerging Intratheater Airlift Demand

Emerging Army future CONOPS emphasize operational maneuver of BCTs (formally designated units of action [UA]) up to 400 km (~215 nm) deep into enemy territory to seize key objectives:

> This concept would allow the UA [now BCT] to be inserted into an OA [operating area] and be supported with some portion (and in surges, possibly all) of their supplies by air. This is necessary for two reasons. First, the UA [BCT] will at times operate in areas without secure ground lines of communication (LOC). Second, the UA [BCT] must be able to move a distance of 400 km upon insertion to the OA. . . . The UA is projected to have a battlefield footprint of approximately 75 × 75 km, and it will operate within a 500 × 500 km area of operations with perhaps up to 5 other UAs.[2]

Over the past few years, future Army CONOPS have matured, and more details have been released. Overall, they increase emphasis on operational maneuver to achieve the joint force commander's campaign objectives. Training and Doctrine Command (TRADOC) Pamphlet 525-3-1 summarizes the overall concept for 2015–2024 as follows:

> In accordance with the joint campaign plan and other components of the joint force, the future Modular Force will ***seize the initiative*** through shaping and entry operations. ***Intratheater operational maneuver*** by ground, air, and sea of powerful, modular, combined arms formations extends the reach of the JFC [joint force commander], expands capability to exploit opportunity, and generates dislocating and disintegrating effects through the ***direct engagement of decisive points and centers of gravity. Simultaneous, distributed operations*** within a noncontiguous battlefield framework enable the force to act throughout the enemy's dispositions, present multiple dilemmas, and, sequenced over time, achieve operational ends more rapidly. ***Continuous operations*** and ***controlled (high) operational tempo***

[2] See U.S. Army Aviation Center, 2003b, p. 17.

> overwhelm the enemy's ability to respond effectively and support a pace of physical destruction and psychological exhaustion not achievable today.[3] [Emphasis in original.]

This concept has a number of key aspects, but one of the most important is the emphasis on simultaneous, distributed operations by multiple BCTs. In fact, the CONOP states that the future Army will be designed around this concept.

> As a force deliberately designed for decentralized, non-contiguous operations, future Modular Force divisions and corps will be employed in simultaneous operations distributed across the entire JOA [joint operations area].[4]

Simultaneous, distributed operations are further described as follows:

> The non-contiguous operational framework expected to characterize future campaigns will also require conduct of defensive operations that may be either short- or long-term in duration. For example, the exposure of the widely distributed facilities of the joint support structure to attack by unconventional forces, long range fires, aviation and the remnants of enemy forces will present additional requirements for ground defense. Moreover, as ground formations quickly advance to critical objectives throughout the Joint Operations Area (JOA), bypassing some enemy forces and leaving other enemy force remnants intact, commanders will often be required to assume the defensive in specific areas in order to respond to small scale enemy attacks, maintain LOCs, or to isolate force remnants until they can be resolved. Conditions will often dictate that corps and divisions dedicate subordinate forces to defend critical support facilities and vital support operations such as logistical convoys. In situations in which commanders choose not to permanently secure all LOCs and bases, this security requirement will demand new solutions that inte-

---

3 TRADOC Pamphlet 525-3-1, "United States Army Operational Concept for Operational Maneuver 2015–2024," Vers. 1.0, Fort Monroe, Va.: Headquarters, U.S. Army Training and Doctrine Command, October 2, 2006, para. 3-2b.

4 TRADOC Pamphlet 525-3-1, 2006, para. 4-3a.

grate air, electronic, and ground defenses of both stationary and moving "islands of infrastructure" within the JOA.[5]

Future Army operational maneuver CONOPS thus envision U.S. ground forces rapidly attacking or seizing key operational objectives simultaneously in widely separated parts of enemy-controlled territory. They may arrive by air or maneuver on the ground, but they will not generally attempt (or desire or have the capability) to control large areas. Under these conditions, ground LOCs will be subject "to attack by unconventional forces, long range fires, aviation and the remnants of enemy forces." If we take this concept seriously, we must conclude that, in many future conflicts, intratheater airlift will be the only practical means of providing logistic support to a large fraction of U.S. ground maneuver units.

Relying primarily on air resupply puts a premium on the ability of air delivery platforms to operate reliably while very close to dispersed ground combat units. This, in turn, implies that the ability of platforms to operate from austere airfields, or with no airfield at all, will be increasingly important in intratheater airlift operations. Throughout the following analysis, we examine cases in which the intratheater airlift system needs to deliver and extract personnel and supplies down to the battalion level to better understand the magnitude of airlift required by this proposed CONOPS.

Analytical Methodology

We evaluated a series of representative intratheater airlift vignettes to identify how potential emerging needs could widen the capability gap resulting from C-130 retirements. The remainder of this chapter describes the methodology, vignettes, and current intratheater airlift force structure used in this analysis.

As discussed in the FAA,[6] intratheater airlift will support joint land forces, with Army support accounting for the majority of USAF intratheater airlift needs. We therefore focused this analysis on sup-

---

[5] TRADOC Pamphlet 525-3-1, 2006, para. 4-3c.

[6] Orletsky, Rosello, and Stillion, 2011.

porting Army ground combat units. However, other intratheater airlift needs, while relatively small, do exist and are not captured in this analysis. Examples of non-Army intratheater airlift demand include the movement of key replacement parts, munitions, and personnel to or between air bases in theater.

Missions

Intratheater airlift capabilities include providing routine sustainment, TS/MC resupply, and maneuver capabilities to U.S. and allied forces across all operating environments.

Routine Sustainment

As previously discussed, *routine sustainment* is defined as the steady-state logistical flow of required supplies and personnel to deployed units. The consumption rate for many items is generally well understood, so the required routine sustainment can be identified and planned well in advance. These items may consist of water, food, and other items needed to conduct planned operations.

We chose to model units conducting attack operations. During these, units tend to move a great deal but engage in combat infrequently. This type of operation allowed us to capture activities during the most critical phases of major combat operation and is representative of the patrol missions common during postcombat operations. Fuel and water typically constitute the lion's share of routine sustainment demand. The predictable nature of this requirement allows preplanned airlift operations and efficiently loaded airlift sorties. The quantity of supplies and the number of personnel that must be moved by air over time and the number of delivery locations that must be supported drive the ability of the intratheater airlift system to fulfill this requirement.

Time-Sensitive, Mission-Critical Resupply

The capability to provide *TS/MC resupply* is generally reflected by the ability of the airlift system to respond to short-turn taskings for crucial equipment, supplies, and personnel. The requirement for this capability is driven by the need for items (1) with unpredictable consumption rates, (2) that are not kept on hand at every operational location, and (3) are required to maintain operational effectiveness.

The majority of items delivered as TS/MC cargo are spare parts to keep equipment operational and emergency supplies of ammunition or other critical items that have been expended faster than predicted (e.g., medical supplies, fuel). TS/MC resupply may also include the delivery of key personnel with specific skills—perhaps personnel required for equipment repair or to conduct a particular task.

The ability of the intratheater airlift system to fulfill this requirement is driven by the quantity of supplies required over time, the frequency of TS/MC resupply cargo movements, the number of personnel, the number of delivery locations to support, and acceptable delivery times.

Maneuver

The *maneuver* mission area is defined as the ability of the intratheater airlift system to transport combat teams around the battlefield. The maneuver task is associated with the initial deployment, redeployment, and extraction of these teams as required. Maneuver missions may include (but are not limited to)

- transport to mission locations before the mission commences
- transport to a mission in progress
- transport from one mission area to another
- transport following completion of a mission, including moving the mission team and any materiel or personnel acquired during operations (rescued personnel or captured enemy or materiel).

The ability of the intratheater airlift system to fulfill the maneuver requirement is driven by the number of teams and items that must be moved over time; the size of the teams, including the required equipment; and the location to which they must be delivered. The supply requirements of combat teams during ongoing operations may fall into either the routine sustainment or the TS/MC resupply task.

In March 2006, the Air Force Deputy Director for Operational Requirements directed that, for the purposes of this FNA, the maneuver mission would follow the general description in the Force Application Joint Functional Concept.[7] This description "concentrates on capabilities required to effectively apply force against large-scale enemy forces in the 2015 time frame" and is the "integrated use of maneuver and engagement." *Maneuver* is defined as the "movement of forces into and through the battlespace to a position of advantage in order to generate or enable the generation of effects on the enemy."[8] We also received the following guidance:

1. Maneuver will be an important driver of demand for intratheater lift.
2. Because the Army's new maneuver concept is insufficiently mature to be assessed, the current FNA will not assess it.
3. The USAF is committed to assessing it when and if the Army can provide a stable foundation for the analysis.

Therefore, the remainder of the FNA will focus on the routine sustainment and TS/MC resupply missions. It is important to recall that we are analyzing what it would take to support a single operational maneuver involving six BCTs by air once they have been inserted. We have not examined what it would take to insert these forces in the first place.[9] Since we are focusing only on these two potential demands, the overall requirement for intratheater airlift could be higher, if the maneuver mission were adopted.

Modeling Routine Sustainment and Time-Sensitive, Mission-Critical Resupply

We developed a spreadsheet model to determine the number of aircraft required to perform the routine sustainment and the TS/MC resup-

---

[7] U.S. Department of Defense, *Force Application Functional Concept*, Washington, D.C., March 5, 2004c, p. 4.

[8] U.S. Department of Defense, 2004c, p. 10.

[9] For discussions of this topic, see Alan J. Vick, David T. Orletsky, Bruce R. Pirnie, and Seth G. Jones, *The Stryker Brigade Combat Team: Rethinking Strategic Responsiveness and Assessing Deployment Options*, Santa Monica, Calif.: RAND Corporation, MR-1606-AF, 2002, and Mahyar A. Amouzegar, Robert S. Tripp, Ronald G. McGarvey, Edward W. Chan, and C. Robert Roll, Jr., *Supporting Air and Space Expeditionary Forces: Analysis of Combat Support Basing Options*, Santa Monica, Calif.: RAND Corporation, MG-261-AF, 2004.

ply intratheater airlift missions. The routine sustainment intratheater airlift model accounts for such factors as the type, number, and mission of Army ground combat units being supplied; cargo type and amount; aircraft payload and volume constraints; aircraft speed; mission distance; climatic conditions; and various times, including loading, unloading, and maintenance.

Ground Combat Unit Demand for Supplies. Using this model, we first estimated the total daily routine sustainment that infantry, Stryker, and heavy Army BCTs require.[10] Using this demand as a starting point, we next computed the number of C-130 sorties needed each day to support each of these three different types of Army units. We chose to evaluate the ability of the intratheater airlift system to supply quantities of supplies derived from tables supplied by the Army Combined Arms Support Command (CASCOM) for units engaged in attack operations because they best capture the types of activity Army ground combat units engage in during both major combat operations and stability and sustainment operations.[11] In addition, the Army's operational maneuver CONOPS requires units to be able to move rapidly from one objective to another, and the existing attack supply levels best capture these sorts of operations.

For our assessment, we looked at two broad categories of cargo—bulk and liquid. We assumed the bulk cargo would be transported in pallets—specifically, 463L pallets. We began with CASCOM planning factors for the three types of BCTs.

Because we wanted to model supply delivery down to the battalion level but lacked specific brigade consumption factors from CASCOM, we assumed that the brigade headquarters would consume 10 percent of the daily brigade supply requirement and that the three maneuver

---

[10] These three types of BCTs are the only ones the Army plans to field in the 2015 time frame. Future Combat System BCTs will begin to be fielded around 2020, but infantry, Stryker, and heavy BCTs will continue to account for the majority of Army maneuver units until at least 2025.

[11] See the appendix for a discussion of the sustainment requirements.

battalions in the brigade would each consume 30 percent.[12] We further assumed that each brigade headquarters would be colocated with one maneuver battalion.[13] Table 4.1 lists the daily supply requirements for all three BCT types.

Calculating the Number of Aircraft Needed to Meet Demand. For each of the transportation options considered, we had data for the number of pallets that could be transported or estimated the number based on the size of the aircraft's cargo compartment. Then, using average pallet weight data (see the appendix), we verified that the maximum cargo capacity did not exceed the maximum weight for each transportation option. This allowed us to calculate the maximum bulk cargo load for each type of aircraft.

We then calculated the maximum payload each aircraft could carry into each airfield in the two vignettes (described below), accounting both for environmental conditions and for distance from the theater's main operating base. For delivery of water and fuel, we assumed the aircraft were equipped with large tanks, estimating tank weights of approximately 5,000 lbs for a C-130 and 3,000 lbs for a CH-47 using an engineering rule of thumb.[14]

We chose the lower of the maximum payload or the payload constrained by airfield and environmental conditions for each aircraft-airfield pair. This allowed us to calculate the total number of sorties required to supply a particular unit type.

Once we determined the number of sorties of each aircraft type required to provide the routine sustainment mission, we determined the total number of each type of aircraft required according to the dis-

---

[12] We understand this represents an old brigade concept and not the current modular design. This, however, does provide a reasonable example for our analysis because we are attempting to understand the magnitude of the challenges potential concepts may present for intratheater airlift.

[13] Some have pointed out that this will often not be the case. However, the headquarters will be deployed somewhere, and it will require a level of resupply consistent with the one we have assumed. Therefore, this assumption (while not operationally accurate in some cases) allows us to account for the overall logistic support demand.

[14] An alternative would be to consider using blivets—flexible, air-transportable bladder tanks—to transport bulk liquids.

Table 4.1
Daily Unit Supply Requirements

	Stryker Brigade		Infantry Brigade		Heavy Brigade	
Cargo Class	463L Pallets	Tons	463L Pallets	Tons	463L Pallets	Tons
Bulk						
I	5.25	10.79	5.25	9.60	5.00	10.29
II	3.75	3.00	3.25	2.60	3.75	2.80
III (pkg)	1.00	0.90	1.25	0.90	3.25	3.60
IV	16.00	16.60	14.25	14.70	15.25	15.80
V	3.24	12.38	1.75	6.69	5.40	20.64
VI	3.25	3.00	2.75	2.60	3.00	2.80
VII	4.50	4.90	6.00	5.50	14.75	20.20
VIII	0.75	0.30	0.75	0.30	0.75	0.30
IX	12.00	12.30	2.75	2.50	13.25	13.30
Total	49.74	64.17	38.00	45.39	64.40	89.73
Liquid	**Gallons**	**Tons**	**Gallons**	**Tons**	**Gallons**	**Tons**
III (bulk)	28,183	94.41	23,228	77.81	97,885	327.91
Water	19,087	76.35	11,536	46.14	19,087	76.35
Total	47,271	170.76	34,764	123.96	116,972	404.26

NOTE: We were struck by the small amount of class V cargo (ammunition) resulting from the CASCOM analysis. There is large variation in the amount of ammunition expended. In some cases, the entire daily amount of ammunition could be expended in a few minutes of intense combat, while in other cases, very little or no ammunition could be expended during a period of days. We were concerned that, in some cases, the class V daily requirement could exceed the CASCOM planning factor estimate. Since the objective of this FNA is to determine if a capabilities gap exists, we wanted a more-conservative (larger) estimate of class V sustainment for our analysis that would represent days during which ammunition expenditure was relatively large because the level of hostilities was high. See the appendix for more details.

tance from each base to the unit receiving the sustainment; the speed of the aircraft; and the load, unload, and maintenance times. Finally, we incorporated a mission-capable rate to determine the total number of vehicles that must be allocated to the routine sustainment task.

We used a similar approach to identify the total number of aircraft that must be allocated to the TS/MC resupply mission. For this approach, the key measure of effectiveness is time. The amount of time that a unit must wait before receiving a critical item that is not typically included in the daily sustainment package is of major importance. The analytical approach is that these missions will be flown on schedule and that the unit being supplied has the opportunity to place anything required up to its allotment for that sortie. These items could be required to repair critical pieces of equipment or could be emergency items, such as ammunition for units that have unexpectedly depleted their supplies.

To evaluate the number of aircraft required to accomplish the TS/MC resupply mission, we assumed a "direct delivery" system, in which fixed-wing aircraft fly cargo and personnel directly to the airfield nearest the battalion they are supplying. The aircraft then return to the major theater logistics hub.[15] For both the routine sustainment and TS/MC resupply missions, we modeled delivery of supplies in detail down to the battalion level.

Vignettes

Recent experience in Afghanistan and Iraq provides insight into the current operational environment and operational approach. Air Force planners have identified an increased demand for air cargo delivery to dispersed, austere locations in response to enemy attacks on ground LOCs. The increased demands for dispersed, direct delivery over long distances under austere conditions may present a problem for the current intratheater airlift system. While C-130s are well suited to long-haul cargo delivery, they do require a suitable landing surface that may not be (1) secure or (2) close enough to maneuvering ground units. Helicopters are well suited to austere (short and soft) landing surface

[15] We considered evaluating various scheduled theater airlift ("STAR") route options where one aircraft carries cargo to several destinations before returning to a main base, but we found the number of aircraft required to support the direct delivery concept was modest. STAR routes would therefore reduce the number of aircraft only a little; the resource savings would not be enough to overset the costs of the increased delivery times.

operation but lack the speed, range, and payload capabilities of the C-130.

We used vignettes designed to evaluate the size and severity of this potential problem in three ways. The first is an operational-level analysis of the amount of routine sustainment future Army forces must receive. The second is the ability of the intratheater airlift fleet to provide TS/MC resupply. The third focuses directly on the perceived inability of the current set of USAF intratheater aircraft to access short, soft fields.

Overview

To explore how climate and the geographic distribution of airfields could affect intratheater airlift effectiveness, we chose two vignettes from two very different settings. The first vignette is based in Afghanistan, the second vignette in Indonesia.

For both vignettes, we identified the units that must be supplied and the bases from which this resupply could take place. In both cases, we used a CASCOM methodology to identify the quantities of supplies required for routine sustainment of each unit. To ensure the sensitivity of the analysis to key variables, we introduced a number of parametric changes to each set of base-case assumptions. For TS/MC resupply, we considered the ability of USAF intratheater assets to provide resupply of equipment outside the routine daily sustainment. This may include items for which the demand is uncertain and items that are not typically kept on hand at the unit level.

Future Army Operational Maneuver Concept of Operations and the Intratheater Airlift System

As discussed earlier, large multi-BCT forces operating without a ground LOC are not the current Army concept for future operations, but the trend is toward more-dispersed ground-force operations. The Army's future concept of operations focuses on the use of operational maneuver at historically unprecedented levels. Some have discussed concepts to free the maneuver force from a heavy logistical tail and thereby permit very rapid maneuver without the need to secure and maintain long, vulnerable supply lines.

The following discussion examines the possibility of supplying a multi-BCT unit entirely by air. While the objective is to assess the potential airlift demand for such concepts, we also hope to provide a sense of the additional airlift requirements for less-ambitious resupply concepts. Such concepts require light, agile maneuver forces that maintain only a limited amount of daily supplies on hand. Although the Army plans to deploy with 72 hours of consumable supplies, plans call for ground combat units to maintain zero inventory of some types of mission-critical spare parts, whose consumption rates are unpredictable.[16]

The operational maneuver CONOP has several major implications for the logistical supply system. First is the requirement that ground maneuver units operate without depending on fixed logistical bases. For the intratheater airlift system, that means an increased requirement to deliver cargo to dispersed ground combat units without reliance on air bases.

The second major implication for the logistical system is the increased importance of maintaining the delivery schedule. Because future maneuver units will carry only a small amount of sustainment and spares, they will depend more heavily on the certainty of resupply.

The third requirement of this operational concept is the volume of supplies that might be required to supply a large combat force. Although there are no accepted initiatives that would involve a major portion of the Army's force structure, the implications of supplying a larger force are worth considering. Supplying a force consisting of tens of battalions scattered throughout the battlefield by air is qualitatively different from supplying a few special operations teams consisting of a dozen or fewer men each.

In the latter case, one small, light aircraft can deliver a few days of supplies by landing on a road or an improvised landing zone. However, conventional infantry and mechanized combat units of battalion size require a large, constant flow of supplies. The weight and volume of sup-

---

[16] Today, about 30 percent of mission-critical resupply is for items with demands too low to be stocked in each BCT. Items in moderate and high demand will likely be carried in sufficient quantities to meet operational readiness goals. RAND colleague Eric Peltz offered this insight.

plies involved is so large that improvised landing zones and all but the largest roads would rapidly be reduced to a maze of ruts by the volume of fixed-wing traffic required to support even a few days of operation.[17] Although the current fleet of USAF fixed-wing aircraft (C-130s and C-17s) has some capability to deliver cargo to austere and unprepared landing strips, operations involving the amount of supplies required to support battalion-size units require fairly robust infrastructure.

For example, C-130s and C-17s currently require flat, unobstructed areas of approximately one-half mile by 100 ft simply to land.[18] Conducting actual operations—operations involving more than one aircraft on the ground, unloading operations, etc.—requires much more land. It is possible to choose land areas with suitable dimensions, but the number of such sites is highly dependent on geographic location. Many such areas may exist in the desert, but areas of this size may be nonexistent in mountainous, forested, or urban areas.

Even after an area of suitable dimensions is found, surface hardness must be considered. Procedures exist to test the ground hardness. These tests require on-site personnel and must be conducted throughout the landing area to ensure the ground can support the aircraft. Insufficient ground hardness can cause significant damage to the aircraft (including aircraft loss) and injury to the crew. In addition, enemy action could be a major problem. If only a few sites exist that are capable of supporting fixed-wing operations and if these sites are not per-

---

[17] Fixed-wing aircraft can land on patches of unprepared ground, roads, and clearings. However, this ability is highly dependent on aircraft characteristics, aircrew training and proficiency, and ground conditions. Additional factors include aircraft type, size and number of landing gear, tire pressure, and aircraft weight; runway length and width, turnaround space, and apron space to unload aircraft and store supplies; soil strength; and the flatness of the surface (e.g., grade, bumps, tree stumps). In addition, many other variables can limit the ability to operate from austere fields. Frost, for example can reduce subgrade strength. Rain can greatly reduce the runway condition reading of nonpermanent surfaces. These uncertainties have led to a set of engineering procedures requiring significant manpower and equipment to improve landing zones so they can be used for extended periods.

[18] According to the Air Mobility Planning Factors, Air Force Pamphlet 10-1403, the minimum recommended runway dimensions for the C-130 are 3,000 ft long by 60 ft wide. A C-17 requires a runway that is 3,500 ft long by 90 ft wide.

manent and protected, mortars, small arms, land mines, and improvised explosive devices could be devastating.

At the same time, it is clear that supplying rapidly maneuvering future ground forces by air without the dependence on fixed sites requires the use of unimproved landing zones. Given that the Army is traveling light and is depending on timely resupply to maintain combat effectiveness, any disruption could delay operations. It is unlikely that commanders would conduct high-risk operations that are overly dependent on just-in-time sustainment.

Moving large volumes of supplies through unimproved landing zones is best accomplished using large helicopters, such as the U.S. Army's CH-47D. This specific aircraft has a considerable payload but lacks the speed and range of fixed-wing aircraft. It is most effective when used on missions of about 150-nm radius or less.[19]

So, the USAF faces a situation in which the primary fixed-wing intratheater airlift platforms require base infrastructure that may not be available at the right time and place during a rapidly moving ground operation. At the same time, our primary rotary-wing intratheater airlift platform lacks the speed and range to fly effectively from theater logistic bases directly to maneuvering ground combat units. We addressed this dilemma by using C-130s to deliver supplies to a number of airfields (some fairly short, but still capable of supporting significant C-130 operations) within easy CH-47D range of ground maneuver units. We then used CH-47Ds to transport supplies and personnel the final 50 to 150 nm to the ground units.

[19] At distances beyond 150 nm from base, the CH-47 cannot carry a useful payload and return to its base without refueling. If an intermediate refueling point is established, the CH-47 must carry fuel to this location, as well as all the supplies required by the maneuver battalions it is supporting. This increases the number of CH-47s needed. In addition, longer round trips result in longer sorties. This reduces the overall CH-47 sortie rate, further increasing the number of CH-47s required. These factors combine to drive an increasingly steep and unrealistic CH-47 demand for delivering supplies at distances beyond 150 nm from a C-130–capable landing surface.

Table 4.2
Afghanistan Vignette Airfield Locations, Sizes, and Elevations

Name		Type	Latitude (north)	Longitude (east)	Length (ft)	Width (ft)	Elevation (ft)
1	Manas	L1	43.061	74.478	13,780	180	2,058
2	Bagram	L2	34.946	69.265	9,852	180	4,895
3	Mazar-e Sharif	L2	36.707	67.210	10,361	150	1,284
4	Herat	L2	34.210	62.228	8,218	150	3,206
5	Kandahar	L2	31.506	65.848	10,532	148	3,337
6	Bamian	L3	34.809	67.818	8,515	75	8,471
7	Khowst	L3	33.333	69.952	6,735	175	3,756
8	Razer	L3	36.023	70.770	2,858	133	8,289
9	Chaghcharan	L3	34.527	65.272	6,635	90	7,382
10	Farah	L3	32.366	62.166	7,860	60	2,201
11	Qal'eh-ye Now	L3	34.985	63.118	5,305	100	2,998
12	Shindand	L3	33.391	62.261	9,140	160	3,773
13	Bost	L3	31.559	64.364	7,650	148	2,464
14	Oruzgan	L3	32.903	66.631	4,670	180	6,725
15	FOB Rhino	L3	30.493	64.521	6,695	125	3,198
16	Tarin Kowt	L3	32.604	65.866	6,355	110	4,429
17	Andkhoi	L3	36.943	65.207	2,455	60	900
18	Sheberghan	L3	36.751	65.913	8,600	70	1,053
19	Shughnan	L3	37.499	71.507	2,635	100	6,700

Afghanistan Vignette

Table 4.2 lists the name, location, length, width, and elevation of each Afghanistan vignette airfield. Table 4.3 shows the distance between any airfield and all other airfields.

For analysis purposes, we divided the available landing surfaces into three categories. Landing surface category one (L1) consists of major air bases with long, hard-surface runways and significant aircraft parking and servicing infrastructure outside the country in which military operations are ongoing. Landing surface category two (L2) consists of relatively long and wide, hard-surface runways inside the country in which military operations are ongoing. Landing surface category three (L3) consists of generally shorter, narrower airfields that,

Table 4.3
Afghanistan Vignette Airfield Distances (nm)

	Manas	Bagram	Mazar-e Sharif	Herat	Kandahar	Bamian	Khowst	Razer	Chaghcharan	Farah	Qal'eh-ye Now	Shindand	Bost	Oruzgan	FOB Rhino	Tarin Kowt	Andkhoi	Sheberghan
L1 Manas																		
L2 Bagram	545																	
L2 Mazar-e Sharif	508	146																
L2 Herat	782	351	286															
L2 Kandahar	807	269	320	245														
L3 Bamian	585	72	118	279	222													
L3 Khowst	622	103	244	389	235	138												
L3 Razer	457	98	177	434	366	162	167											
L3 Chaghcharan	669	199	162	152	184	127	244	284										
L3 Farah	868	387	361	111	195	319	397	481	203									
L3 Qal'eh-ye Now	718	303	225	64	250	232	354	379	110	164								
L3 Shindand	817	361	315	49	214	289	386	449	165	62	105							

Table 4.3—Continued

	Manas	Bagram	Mazar-e Sharif	Herat	Kandahar	Bamian	Khowst	Razer	Chaghcharan	Farah	Qal'eh-ye Now	Shindand	Bost	Oruzgan	FOB Rhino	Tarin Kowt	Andkhoi	Sheberghan
L3 Bost	842	320	340	192	76	261	303	417	184	122	215	153						
L3 Oruzgan	714	180	230	234	93	129	169	278	119	228	215	222	141					
L3 FOB Rhino	893	359	397	252	92	308	325	457	245	165	279	209	65	181				
L3 Tarin Kowt	749	221	255	206	66	165	211	318	119	188	198	188	99	43	144			
L3 Andkhoi	563	231	97	220	328	181	318	274	145	313	156	258	326	253	389	263		
L3 Sheberghan	547	196	63	236	315	149	286	239	137	322	173	270	321	234	383	249	36	
L3 Shughnan	361	188	211	493	456	241	262	96	352	554	434	515	502	365	546	404	303	272

in many cases, have unpaved, austere landing surfaces inside the country. In general we assume that supplies flow via strategic lift into L1 airfields. From there, intratheater airlift platforms deliver them to L2 and L3 airfields. We assume BCT headquarters are usually located, along with one maneuver battalion, at or near an L2 airfield, with the remaining battalions located at or near L3 airfields.

For the Afghanistan vignette, we identified one large "main operating base," Manas. Manas is outside Afghanistan, and we assumed it to be the terminus of strategic airlift operations. In addition, we identified four bases inside Afghanistan that would be used as forward bases from which to provide the sustainment and resupply to the deployed units. We classified these as L2 bases. All four are capable of significant logistic throughput. Finally, we identified 14 more-austere operat-

Figure 4.1
Afghanistan Vignette Airfield Map

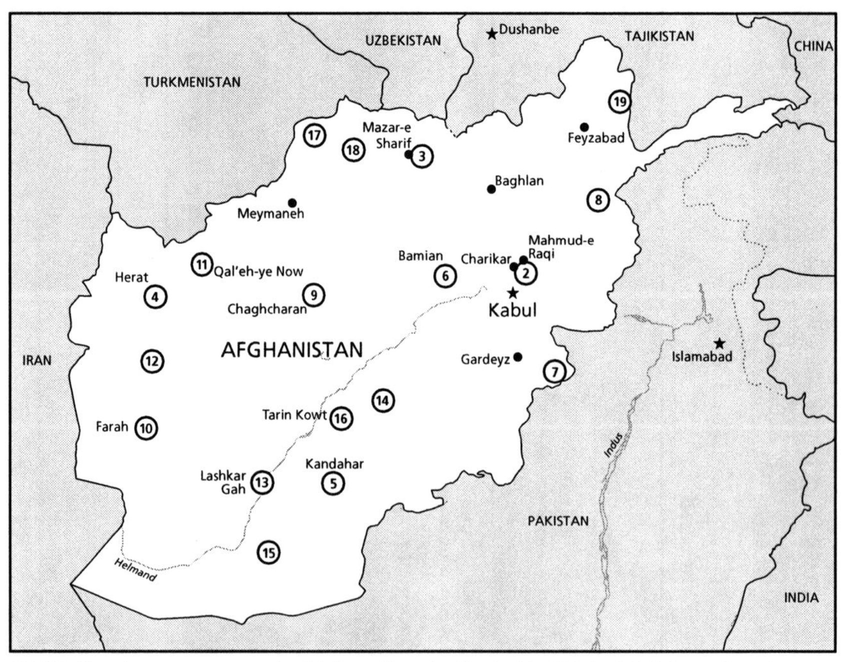

NOTE: Numbers correspond with locations indicated in Table 4.2; Manas is not within the map area.
RAND *MG822-4.1*

ing locations inside Afghanistan that are capable of supporting C-130 operations. For convenience we refer to these as L3 airfields or forward operating locations (FOLs). Figure 4.1 shows the locations of all L2 and L3 airfields.

The baseline Afghanistan vignette evaluates the ability of the current force of C-130s to completely supply four infantry BCTs and two Stryker BCTs. This six-BCT force is a stylized case; other units would also be deployed. Here, it serves as an example of the potential resupply requirement. We placed these brigades around the country as our base case. Figure 4.1 shows our base case locations for these units, and Table 4.1 presents the daily sustainment for each of these units. The demand for supplies shown in this table represents the routine daily sustainment in terms of both short tons and 463L pallets.

Operations in Afghanistan have several distinct characteristics that make this vignette interesting and challenging. First, Afghanistan is highly mountainous. Operations often take place at relatively high altitudes that adversely affect aircraft performance—especially helicopter and fixed-wing aircraft takeoff performance. In fact, the average elevation of the L2 and L3 airfields shown in Figure 4.1 is almost 4,200 ft above mean sea level, with five of the 18 L2 and L3 airfields at or above 6,500 ft. In addition, Afghanistan is a desert country, and summertime temperatures can be quite hot. The combination of high altitudes and hot temperatures results in a very challenging aircraft operating environment. For this analysis, we assumed that all aircraft operations in Afghanistan occurred on "hot" days, which we defined as 108°F at sea level.[20]

Second, Afghanistan is rugged, with limited and greatly deteriorated transportation infrastructure (roads and airports). The combination of a high, hot operating environment and poor (generally short and soft) airfield infrastructure is a challenging test case for the ability of the existing intratheater airlift system to fully supply ground combat unit needs.

[20] To calculate aircraft performance flying into and out of the various airfields, we adjusted the temperature to match the airfield elevation using a standard temperature lapse rate table.

Table 4.4
Indonesia Vignette Airfield Locations, Sizes, and Elevations

	Name	Type	Latitude (+north, –south)	Longitude (east)	Length (ft)	Width (ft)	Elevation (ft)
1	Paya Lebar	L1	1.360	103.910	12,401	200	65
2	Tindal	L1	–14.521	132.378	9,003	150	443
3	Iswahyudi	L2	–7.616	111.434	8,448	92	361
4	Moses Kilangin	L2	–4.528	136.887	7,841	148	103
5	Mutiara	L2	–0.919	119.910	6,781	98	284
6	Tjilik Riwut	L2	–2.225	113.943	6,890	98	82
7	Batujajar	L3	–6.904	107.476	5,420	65	2,500
8	Gorda	L3	–6.140	106.344	5,250	330	40
9	Tambolaka	L3	–8.597	120.477	5,905	98	161
10	Bali International	L3	–8.748	115.167	9,790	148	14
11	Dominique Edward Osok	L3	–0.895	131.285	6,070	90	10
12	Kaimana	L3	–3.645	133.696	5,249	98	19
13	Mopah	L3	–8.520	140.418	6,070	98	10
14	El Tari	L3	–10.172	123.671	4,175	210	335
15	Jalaluddin	L3	0.637	122.850	7,407	100	105
16	Presidente Nicolau Lobato International	L3	–8.547	125.525	6,065	98	25
17	Sam Ratulangi	L3	1.549	124.926	8,693	148	264
18	Pinang Kampai	L3	1.609	101.434	5,905	98	55
19	Sultan Iskandarmuda	L3	5.524	95.420	8,184	148	65
20	Pangsuma	L3	0.836	112.937	3,294	75	297

Indonesia Vignette

Table 4.4 lists the Indonesia vignette airfields along with their locations and other characteristics, and Figure 4.2 maps the airfields. Table 4.5 shows the distance between any airfield and all other airfields for the Indonesia vignette.

A number of important variables could affect the demand for aerial resupply of ground units related to climate and geography. With the limited time and resources available to us, we decided to examine

Figure 4.2
Indonesia Vignette Airfield Map

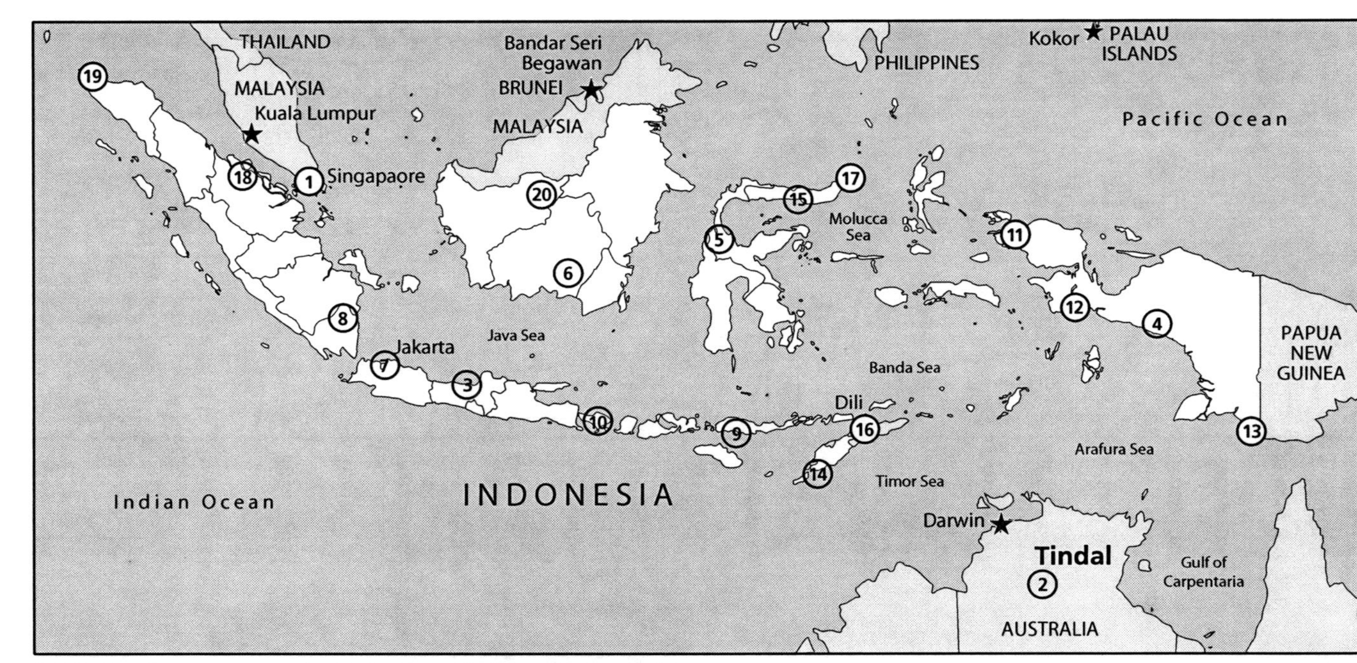

NOTE: Numbers correspond with locations indicated in Table 4.4.

RAND *MG822-4.2*

Table 4.5
Indonesia Vignette Airfield Distances (nm)

		Paya Lebar	Tindal	Iswahyudi	Moses Kilangin	Mutiara	Tjilik Riwut	Batujajar	Gorda	Tambolaka	Bali Intl	Dominique Edward Osok	Kaimana	Mopah	El Tari	Julaluddin	Presidente Nicolau Lobato Intl	Sam Ratulangi	Pinang Kampai	Sultan Iskandarmuda
L1	Tindal	1,944																		
L2	Iswahyudi	703	1,302																	
L2	Moses Kilangin	2,012	657	1,532																
L2	Mutiara	971	1,103	648	1,042															
L2	Tjilik Riwut	640	1,320	357	1,384	367														
L3	Batujajar	541	1,539	240	1,765	828	479													
L3	Gorda	474	1,618	316	1,830	872	513	82												
L3	Tambolaka	1,159	786	541	1,010	463	547	781	855											
L3	Bali Intl	908	1,070	232	1,321	550	399	471	549	316										
L3	Dominique Edward Osok	1,651	822	1,256	401	684	1,045	1,472	1,529	796	1,074									
L3	Kaimana	1,815	658	1,352	199	844	1,189	1,581	1,645	844	1,148	220								

Table 4.5—Continued

		Paya Lebar	Tindal	Iswahyudi	Moses Kilangin	Mutiara	Tjilik Riwut	Batujajar	Gorda	Tambolaka	Bali Intl	Dominique Edward Osok	Kaimana	Mopah	El Tari	Julaluddin	Presidente Nicolau Lobato Intl	Sam Ratulangi	Pinang Kampai	Sultan Iskandarmuda
L3	Mopah	2,266	595	1,725	319	1,310	1,628	1,964	2,036	1,185	1,500	713	497							
L3	El Tari	1,370	574	743	857	600	752	982	1,059	212	511	720	715	998						
L3	Jalaluddin	1,139	1,073	845	898	200	562	1,027	1,071	573	728	515	701	1,188	652					
L3	Presidente Nicolau Lobato Intl	1,425	540	840	720	568	790	1,079	1,152	300	616	575	570	885	147	575				
L3	Sam Ratulangi	1,263	1,063	979	806	336	698	1,164	1,207	666	851	409	612	1,108	709	136	608			
L3	Pinang Kampai	150	2,081	817	2,161	1,121	786	627	551	1,295	1,032	1,800	1,964	2,414	1,507	1,288	1,567	1,412		
L3	Sultan Iskandarmuda	567	2,511	1,243	2,561	1,520	1,206	1,040	960	1,725	1,462	2,187	2,363	2,826	1,937	1,672	1,993	1,786	431	
L3	Pangsuma	543	1,480	516	1,474	432	194	569	576	725	591	1,108	1,276	1,740	922	596	942	722	693	1,088

what we consider to be a worst-case example of the effects of geography and climate on aerial resupply demand. In thinking about how geography and climate might increase aerial resupply demand, however, we determined that operations in a nation with little or no ground supply-route infrastructure, a hot climate, and greater distances than in the Afghanistan example would represent a very stressing case. For illustrative purposes, we chose to examine how the requirements for supporting Army maneuver forces by air would change if the forces were deployed to Indonesia rather than Afghanistan.[21]

In contrast with Afghanistan, Indonesia is composed of thousands of islands. Much of the land area is covered with tropical forest. In addition, Indonesia is very large, so the distances intratheater airlift platforms might be required to fly are considerably longer than in Afghanistan.

The major difference between the two vignettes is that, while the Afghanistan scenario stresses aircraft performance at high altitudes and hot conditions, the Indonesia scenario stresses long-range performance into airfields relatively close to sea level in a tropical environment. We assumed a "tropical" day, with a temperature of 90°F at sea level. As in the Afghanistan case, we adjusted temperature to compensate for airfield elevation. However, unlike in Afghanistan, airfields in Indonesia tend to be relatively close to sea level. Our selected group of L2 and L3 airfields has an average elevation of only 262 ft above mean sea level and an average length of over 6,500 ft. Only one of the airfields is above 1,000 ft elevation at the still-modest altitude of 2,500 ft.

However, because Indonesia is so much larger than Afghanistan, the average distance between L3 airfields is about 3.5 times as far.

---

[21] We do not believe it is either likely that large U.S. ground units will be deployed or desirable that they be deployed to Indonesia. We cannot imagine a situation that would require this. However, the intratheater airlift recapitalization decisions made over the next few years will almost certainly define this capability for the first half of this century. Therefore, it is prudent to explore demanding cases even if they seem unlikely—as the idea of simultaneous U.S. occupations of Iraq and Afghanistan would have seemed to defense planners in the 1960s and 1970s, when our current intratheater airlift capability was envisioned.

Methodology for Computing C-130 and CH-47 Performance

The current fleet providing sustainment and resupply services to Army forces consists of Army CH-47D and USAF C-130 aircraft.

The C-130H

The C-130 is capable of landing on short, austere airfields. We used Portable Flight Planning Software (PFPS) version 3.3.1 to perform detailed calculations for the C-130H aircraft, the most numerous USAF variant of the C-130.[22] This program calculates the minimum field length for takeoff and landing ground roll for the aircraft from gross weight, air temperature, and airfield elevation. In accordance with USAF regulations, we used the greater of minimum field length for maximum-effort takeoff or landing ground roll plus 500 ft to determine the maximum weight at which a C-130 could operate from a given airfield under our assumed climatic conditions.

Given the maximum C-130 operating weight for a given field, we calculated the amount of fuel it would require to get from a given L1 base to a given L2 or L3 base and back without refueling. We then added a standard reserve of 7,000 lbs of fuel to calculate a total fuel weight for the sortie. Subtracting this and a standard operational empty weight of 88,000 lbs from the maximum operating weight for each airfield gave the maximum payload a C-130 could carry into any given airfield. As an additional constraint, we required any C-130 flying into an airfield shorter than 4,000 ft to land no heavier than the maximum allowable assault landing weight of 130,000 lbs.

In every case, the C-130 was able to deliver a full load of six pallets because the low density of palletized cargo ensured that, at all but the shortest and most distant fields, the allowable payload weight was greater than the weight of six pallets. However, for liquid cargo, the payload weight, rather than size restrictions, limited the amount of fuel and water the C-130s could deliver in a much larger number of cases.

---

[22] PFPS is an official USAF model used by aircrews that is based on the aircraft performance manual.

The calculations for Andkhoi Afghanistan are an example. The airfield is 563 nm from the L1 base. The field is at an elevation of 900 ft and is only 2,455 ft long, so in this case, the maximum aircraft landing weight is 130,000 lbs for an assault landing with maximum effort braking.

Subtracting the 7,000-lb fuel reserve from this maximum landing weight leaves 123,000 lbs. We must also subtract the 9,706 lbs of fuel required to fly back to the L1 base.[23] This leaves 113,294 lbs. Subtracting the operational empty weight of the aircraft, about 88,000 lbs, from this leaves a maximum payload of 24,294 lbs. Although this is only about 60 percent of the C-130H maximum payload, it is still quite a respectable performance.

Once we knew the maximum payload weight for each airfield, we could calculate the number of pallets each C-130H can carry (in all our cases, the maximum of six) and the amount of fuel and water each aircraft can deliver at its maximum landing weight. With these numbers in hand, we calculated the total number of C-130 round trips required to supply each type of unit.

The CH-47D

The CH-47D is the backbone of the Army's heavy-lift helicopter fleet. Almost all CH-47D aircraft are conversions of earlier versions of the CH-47. The first of a total of 472 conversions was completed in 1979, and the CH-47 entered operational service in 1984.

The CH-47D has an empty weight of approximately 23,400 lbs and a maximum gross takeoff weight of 50,000 lbs. Its normal internal fuel load is 6,600 lbs, which gives a range of about 230 nm at sea level with maximum payload. The maximum payload weight is 20,000 lbs. The CH-47D's cargo bay is 6.5 ft high, 7.5 ft wide, and 30 ft long, compared to the C-130H's, which is 10 ft high, 10 ft wide, and 40 ft long.[24] This gives the CH-47 half the maximum payload weight capability of a C-130H but only about 37 percent of the cargo volume of

[23] Assuming that no refueling capability exists at L3 class bases.

[24] See GlobalSecurity, "CH-47D Chinook," Web page, September 22, 2005, for more details.

a C-130H. Because most classes of supply fill aircraft volumes before reaching their maximum payload weight capabilities, the number of CH-47 sorties required to distribute supplies is generally limited by cabin volume. These limitations are balanced by the aircraft's capability to land in areas not much larger than the helicopter itself.

Intratheater Airlift Fleet Sizes

Finally, it is worth noting the number of C-130 and CH-47D aircraft the Air Force and Army have available to conduct operational airlift missions.[25] Table 4.6 breaks down the USAF C-130 PMAI by aircraft subtype and component (active duty or reserve).

The table shows that the USAF has a total of 368 C-130s available for operational airlift missions. A large fraction of the USAF C-130s are the C-130H series, which was the aircraft modeled in detail here. It is worth noting that the C-130Es (primarily the active-duty aircraft)

Table 4.6
USAF C-130 Primary Mission Aircraft Inventory

PMAI	Total	Active	Reserve
C-130E	92	65	27
C-130H	252	58	194
C-130J	24	0	24
Total	368	123	245

SOURCE: C-130 inventory tracking spreadsheet maintained by AMC/A8PF. Current as of April 2006. In addition to the PMAI aircraft there were a total of 55 C-130s assigned to training missions for a total of 413 operational and training aircraft. The reader will note that these are PMAI aircraft, rather than the TAI discussed earlier. In addition to PMAI, TAI includes training, development, and other nonoperational aircraft, as well as backup aircraft inventory and attrition reserve.

---

[25] The aircraft available to conduct operational airlift missions are known as *primary mission aircraft inventory* (PMAI). These are aircraft that belong to operational units, as opposed to units devoted strictly to training or to backup inventory.

and the older C-130Hs (active-duty C-130H1s[26]) have experienced age-related CWB problems sooner than anticipated, which may be the result of higher-than-anticipated use rates or structural deficiencies.[27] For safety reasons, many are under restrictions that limit the amount of payload they can carry, and the Air Force is considering retiring them within the next ten years. Therefore, in our discussion of our results in the next chapter, we will look at three different levels of C-130 capability.

The first case posited the availability of the full 368 PMAI C-130s. This can be thought of as the capability of the C-130E fleet having either been restored to full usefulness or been replaced. In the second, we assumed retirement of the C-130Es without replacement, leaving the USAF with only 276 PMAI C-130s. In the third, we assumed that 80 percent of the active aircraft and 25 percent of the reserve aircraft would be available.

---

[26] The C-130H1s are the oldest C-130Hs.

[27] Email communication with a representative of Air Mobility Command's Force Structure Branch.

CHAPTER FIVE

Analysis Results

Both the routine sustainment and TS/MC resupply missions would be crucial for implementing an air-centric theater logistics system, and their requirements would be additive. This chapter presents the results of our analysis of the relevant issues, evaluating the capabilities of the current intratheater airlift system to perform these missions along two key dimensions. The first is the ability of the system to provide the necessary volume of supplies to ground combat units. This dimension is more critical for the routine sustainment mission than for the TS/MC resupply mission. The second is timeliness. Analytically, this dimension is less critical for the routine sustainment mission because of its predictable nature. However, it is extremely important for evaluating the TS/MC resupply needs.

For both dimensions, we conducted parametric sensitivity analysis to show how such variables as distance, supply quantity, time sensitivity, and airfield characteristics affect the number of aircraft required and the potential for capabilities gaps. This parametric evaluation includes the number of aircraft required to perform the routine sustainment and TS/MC resupply missions as a function of resupply distance for each of the units considered (infantry, Stryker, and heavy brigades). We then present the results in the context of the vignettes described in the previous chapter.

We focus first on routine sustainment not because it is more important but because it requires far more resources than TS/MC resupply. The key metric here is supplying the required daily volume of supplies necessary to sustain ground-force operations, and the analysis focuses on this metric. In addition, the section describes a number of sensitivity cases and presents related results.

Routine Sustainment

The most important factor for successful routine sustainment is to move the required volume of supplies forward at the required daily rate. So, for this mission, the key metric is the number of existing platforms required to supply the units in our vignettes with their estimated daily sustainment needs.

Volume and Number of Aircraft Required

Delivering 100 percent of the routine sustainment supplies (see Table 4.1) down to the battalion level for a six-BCT ground force is a major effort.[1] A number of variables can affect the number of intratheater airlift assets required to accomplish this task. First is the mix of BCT types that make up the ground force. To explore how this might affect the overall level of intratheater airlift capability needed to supply the ground force, we looked at two different BCT mixes. We first assumed that four of the six deployed BCTs were light infantry BCTs and that two were Stryker BCTs—we call this our "light BCT" mix. Recall from Table 4.1 that infantry BCTs require less bulk, water, and fuel supplies each day than a Stryker BCT. Our second ground force BCT mix has three infantry, one Stryker, and two heavy BCTs—we call this our "heavy BCT" mix. Again, recall from Table 4.1 that the heavy BCT requires much more ammunition and fuel than a Stryker

---

[1] Again, there are no current plans to supply 100 percent of the daily sustainment for a force of this size by air. Further, current plans call for over 40 active-duty Army BCTs, with an additional 30 in the National Guard. Therefore, the analysis presented here examines what it would take to supply 100 percent of the logistic requirements to less than 10 percent of planned Army forces. This is nearly equivalent to supplying one-third of the logistics demand of the 18 BCTs deployed in Iraq. The case of providing 100 percent of the sustainment of the six-BCT force here is an example of how providing significantly more aerial resupply than is currently done can affect the intratheater airlift fleet. We intend this example to suggest how different operational and sustainment concepts can affect the intratheater airlift fleet.

BCT. So, by replacing one infantry and one Stryker BCT with heavy BCTs we vary aggregate demand for supplies quite a bit.

A second key ground force variable that can affect the amount of intratheater airlift capability needed is whether the organic firepower of the BCTs is augmented with an indirect fires element and an attack helicopter capability.[2] To evaluate how such augmented BCTs would affect the intratheater airlift system, we estimated the additional supplies the augmenting artillery and Army aviation elements would need each day (see the appendix for details) and then used our model to estimate the amount of C-130 capability that would be required for both our light and heavy BCT mixes. We refer to BCTs accompanied by artillery and attack aviation as our "augmented fires" case.

In both the unaugmented and augmented fires cases, we assumed that each of the three battalions of each BCT centers its operating area on one of our C-130–capable airfields. Therefore, it is only necessary for the intratheater airlift system to deliver supplies to the airfield. However, ongoing discussions between the USAF and Army have revealed that supplies must ultimately be delivered to a well-defined area within about 3 nm of each battalion operating location. It is, however, unrealistic to expect that a C-130–capable airfield will always be within 3 nm of each battalion operating area, so another way to deliver supplies to the battalions from the C-130 capable airfields must be found. Since our goal was to eliminate as much ground-supply convoy traffic as possible, we focused on how the Army's fleet of CH-47D helicopters could accomplish this task.

In what we refer to as the "augmented fires with helicopters" case, we analyzed how much intratheater airlift capability would be required to support a scheme in which helicopters deliver supplies from our C-130 capable airfields to battalion operating areas. In the Afghanistan vignette, the greatest distance from any airfield to its closest neighboring airfield was 110 nm, meaning that a C-130–capable airfield would always be within 55 nm of any point within that operating area. Therefore, we assessed a conservative case in which all CH-47D mis-

---

[2] There would also be several other unit types. We focused on fire support and attack helicopters as an example.

sions deliver supplies to battalions operating 55 nm from a C-130 FOL. In the Indonesia vignette, the distance to the nearest C-130 FOL was about four times farther than in the Afghanistan vignette because the country is so much larger. This means that, to support ground unit operations in almost any part of the operating area in Indonesia, the CH-47Ds would need to fly missions with radii of around 200 nm or more. The CH-47D is not capable of delivering useful loads at that distance with enough fuel to return to base; the limit of its useful cargo-delivery radius is about 150 nm.

Attempting to extend this radius by using CH-47s to move fuel to forward refueling points exponentially increases the already large CH-47 tail requirement. Therefore, we limited further analysis of CH-47D sorties in the Indonesia vignette to 150-nm radius missions. This led to the important conclusion that, in large countries with sparse airfield infrastructure, significant areas are simply beyond the reach of the existing intratheater airlift fleet to supply Army maneuver units. In the Indonesia case, the CH-47's range limitation, combined with the sparse airfield availability, means that the overall ability of the intratheater airlift fleet to support Army combat units covers only about 55 percent of the country.

We modeled the number of CH-47D trips required to move all supplies to battalions operating far from C-130–capable airfields. We then calculated the amount of fuel the CH-47Ds would burn on these trips, added this to the daily routine sustainment fuel requirement for each unit, and adjusted the number of C-130 trips accordingly. Table 5.1 shows the results of our analysis.[3]

---

[3] To determine the duration of each sortie, we made the following assumptions: 1 hour to load and unload the aircraft and 1.5 hours of maintenance time per round trip conducted at the main base. The flight time is based on the distance from the main base to each operating base at 300 kts with an additional half hour to account for climb and decent time, which are not accounted for by using an average cruise speed. Finally, we assumed an 85-percent mission capable rate for the C-130H.

We computed the CH-47D payload much as we did the C-130 payload and considering both liquid and bulk sorties. We assumed one 463L pallet per CH-47D sortie. Instead of computing times for each component of the CH-47D, we assumed five sorties (2.5 round trips) per day because none of the sorties involved more than 2.5 hours of flight time round trip. This assumption is consistent with Army CH-47 planning factors.

Table 5.1
Intratheater Airlift Analysis Results

Case	Aircraft	Metric	BCT Mix Light	BCT Mix Heavy
Afghanistan vignette				
Unaugmented	C-130	Round trips	108	162
		Aircraft, planning factor	74	113
		Aircraft, non–planning factor	40	61
Augmented fires		Round trips	258	346
		Aircraft, planning factor	175	240
		Aircraft, non–planning factor	96	130
Augmented fires, with helicopters		Round trips	330	433
		Aircraft, planning factor	221	298
		Aircraft, non–planning factor	122	162
	CH-47D	Round trips	1,056	1,332
		Aircraft	423	533
Indonesia vignette				
Unaugmented	C-130	Round trips	110	161
		Aircraft, planning factor	98	144
		Aircraft, non–planning factor	48	70
Augmented fires		Round trips	263	342
		Aircraft, planning factor	235	306
		Aircraft, non–planning factor	114	149
Augmented fires, with helicopters	C-130	Round trips	496	626
		Aircraft, planning factor	439	557
		Aircraft, non–planning factor	214	271
	CH-47D	Round trips	1,191	1,522
		Aircraft	476	609

The entries in the third column of Table 5.1 can be decoded as follows: *round trips* refers to the total number of trips from the L1 base to the L2 or L3 base location and back that must be flown each day to supply the units involved. *Aircraft* refers to the total number of aircraft that must be deployed to fly the number of round trips listed each day. Air Force Pamphlet 10-1403, "Air Mobility Planning Factors," states that, for planning purposes, it is best to assume that a C-130 engaged in intratheater airlift operations spends just 6 hours each day in the air. If we use this constraint, the number of C-130s required is that listed in the planning-factor row. However, our modeling results indicate that it should be possible to fly each C-130 substantially more each day. USAF Air Mobility Command uses myriad variables to calculate wartime planning factors. These variables include crew and maintenance ratios, crew duty day, aircraft turn time, fleet reliability, spare parts, and funding. For this FNA, we did not assess ways to increase wartime planning-factor utilization rates. "Non–planning factor" is the number of C-130s required if we do not limit the aircraft to 6 hours of flight time per day but fly the C-130s to the limit of their capability under our ground-time and mission-capable rate assumptions.

Figures 5.1 through 5.5 graph data very similar to those in Table 5.1. The figures differ in that the total number of PMAI C-130 or CH-47 aircraft available to the Air Force and Army are depicted as lines across the charts. For the C-130 charts, the top line shows the total PMAI, while the bottom line includes the notation "no Es." This notation means that the size of the C-130 PMAI fleet has been reduced by 92 (to 276) to reflect the retirement of the C-130E model aircraft.

The results are quite interesting. First, averaging the number of C-130 tails required in all light-force cases and comparing it to an average for all heavy-force cases showed that, on average, about 38 percent more C-130s were required to support our heavy force mix than the light force mix. The number of CH-47Ds also increased for the heavy force, but only by about 28 percent.[4]

Second, Figure 5.1 shows that the light force mix in the Afghanistan vignette never exceeds the capacity of the fixed-wing intratheater airlift fleet to supply it. This was true even when we assumed the

[4] We assumed a CH-47D PMAI force structure of 298 aircraft, given input from Headquarters U.S. Army staff.

Figure 5.1
C-130 Aircraft Required for the Afghanistan Light Force Mix

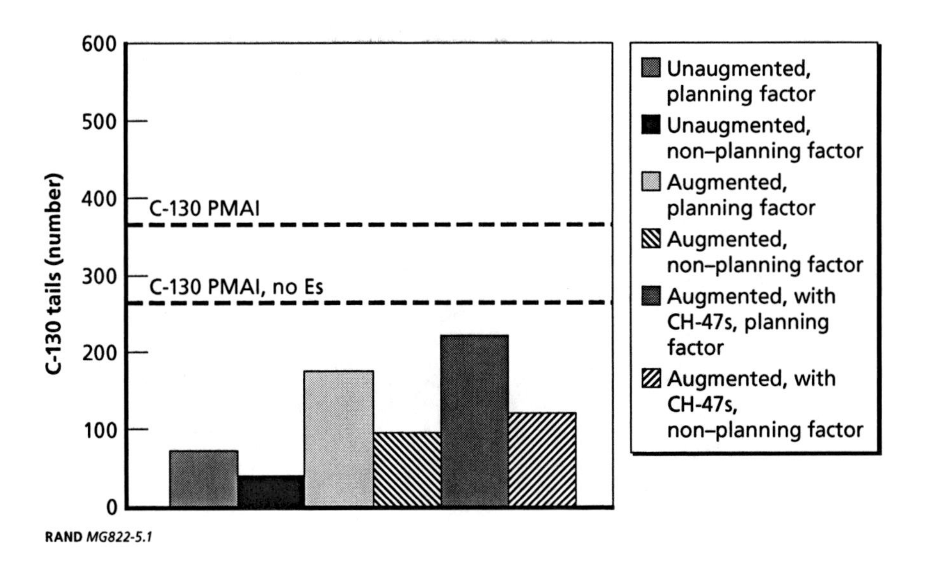

RAND *MG822-5.1*

Figure 5.2
C-130 Aircraft Required for the Afghanistan Heavy Force Mix

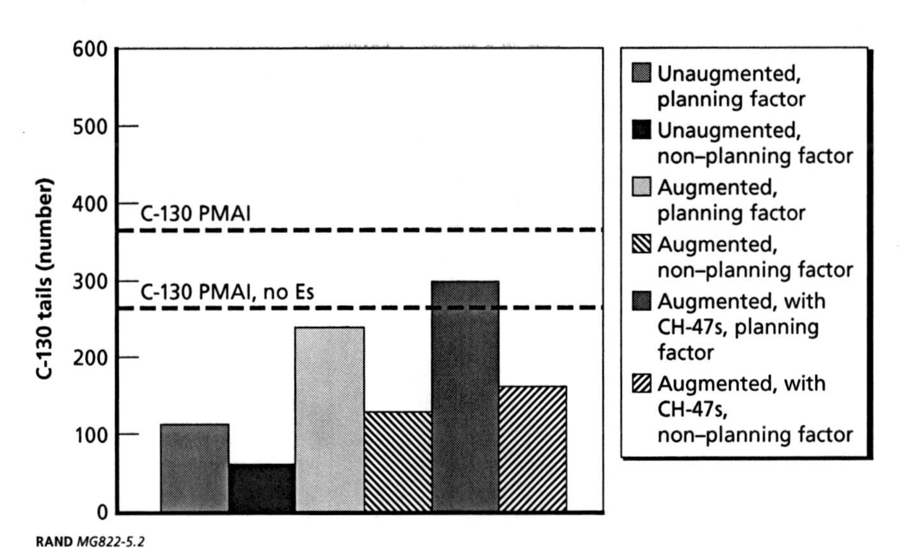

RAND *MG822-5.2*

Figure 5.3
C-130 Aircraft Required for the Indonesia Light Force Mix

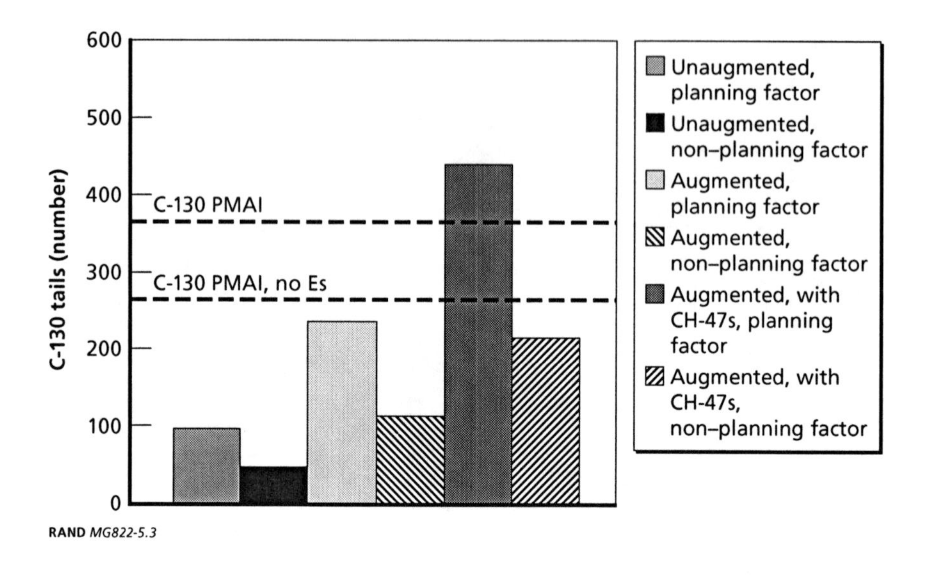

RAND *MG822-5.3*

Figure 5.4
C-130 Aircraft Required for the Indonesia Heavy Force Mix

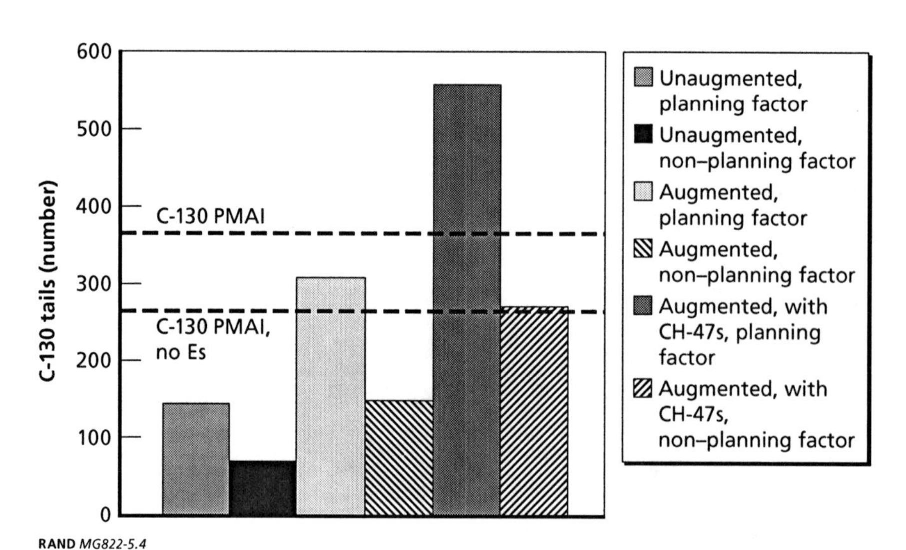

RAND *MG822-5.4*

Figure 5.5
CH-47 Aircraft Required for the Indonesia Heavy Force Mix

RAND *MG822-5.5*

C-130Es were retired and not replaced and when augmented units were deployed and had to be supplied by helicopter. However, even this force mix exceeded the capability of the CH-47 fleet to move sustainment supplies forward if it were augmented with fires and aviation assets.

Third, the difference between the Afghanistan and Indonesia cases was significant for the less-realistic unaugmented and augmented fires cases but was quite large for our most realistic augmented fires with helicopters case.[5] Analysis of the numbers in Table 5.1 shows that the augmented fires with helicopters case in Indonesia required 45 to 50 percent more C-130 round trips than the Afghanistan case. In addition, because the distances involved were longer, the number of C-130s required increases by an average of about 36 percent (depending on PMAI assumptions).

Fourth, the augmented fires with helicopters case—in addition to being the most realistic—was by far the most demanding. The need to

---

[5] We consider this case the most realistic because it is likely that U.S. Army forces inserted deep into enemy territory would be augmented with the fires and aviation assets to increase their effectiveness and survivability.

transport fuel for the helicopter missions greatly increased the number of C-130 tails required. Of particular interest is the difficulty the intratheater airlift system would likely have supplying our heavy force mix in this case. Even in the Afghanistan vignette, the number of C-130s required using standard planning factors was more than two-thirds of the entire existing C-130 PMAI capability, exceeding the capacity of the C-130 fleet if the C-130Es were retired and not replaced. In addition, the heavy force mix with augmented fires required about 60 percent more CH-47Ds to supply it than currently exist.

In the Indonesia vignette, even the light force mix would be impossible to supply with the existing C-130 fleet using standard planning factors and would require 160 percent of the Army's CH-47D fleet to supply in the augmented fires with helicopters case. The heavy force mix would be even more difficult to support. Even with the most favorable assumptions—no C-130Es retire and average daily flight hours increase greatly—the heavy force mix would require 74 percent of the C-130 fleet to supply its daily sustainment needs. Under any other set of assumptions the demand for supplies exceeds the capacity of the C-130 fleet to meet it, and the requirement for CH-47D aircraft exceeds the current fleet size by more than 100 percent.

It is important to recall that we were assessing what it would take to support a single operational maneuver involving six BCTs by air once they had been inserted. We have not examined what it would take to insert these forces in the first place. For perspective, the U.S. Army's total force currently consists of more than 70 BCTs. So the analysis presented so far shows that providing supplies by air to less than 10 percent of U.S. Army BCTs is extremely challenging and, in many cases, is beyond the capability of existing intratheater airlift assets.

The discussion up to this point has assumed the availability of 100 percent of the PMAI C-130 and CH-47D fleets to support either vignette. It is highly unlikely that this level of capability will be available for any single operation. As of this writing, the United States is currently involved in two major operations (Iraq and Afghanistan). In addition, other demands exist for these aircraft in both the United States and overseas. Further, deploying 100 percent of the force is unsustainable for an ongoing operation. This leads to the conclusion that the current intratheater airlift fleet cannot be expected to sustain more than a very small fraction of the Army's planned forces in combat.

Potential Capability Gaps

The preceding discussion highlighted several potential intratheater airlift capability gaps. First, if we assume the entire PMAI is available, the current C-130 fleet seems to be adequate to support either of our force mixes in a country the size of Afghanistan. However, its ability to support more than one such operation is questionable—especially if the C-130Es are retired and if their capability is not replaced. In addition, it is unrealistic to assume that such large fractions of the existing C-130 capability would be committed to any one operation, given the diverse demands of such other missions as supporting the global war on terror, homeland security, and disaster relief.

More-realistic assumptions about the amount of intratheater airlift capability likely to be committed to a single operation indicate that the C-130 fleet is inadequate for supporting free-ranging ground operations by augmented forces unless a light force mix is deployed and unless the current C-130 planning factor utilization can be dramatically improved upon.

The Indonesia vignette results show that, in a country of this size, even the entire current C-130 fleet is unable to support either force mix using standard planning factor assumptions. Supporting unaugmented forces might be possible but only over about 55 percent of the country. However, even if the C-130 fleet could support the light force mix in the Indonesia vignette, it would require more than the entire Army CH-47D force structure to deliver the supplies to maneuvering ground combat units. Adding heavy ground combat units to the Indonesia vignette places great strain on the entire C-130 fleet. Even under our most optimistic set of assumptions, we estimate that it would require 74 percent of the existing C-130 fleet to supply the heavy force mix. Little or no C-130 capability would be left over to support other operations anywhere else in the world. In addition, the entire Army CH-47D fleet is inadequate to supply the heavy force mix.

Time-Sensitive, Mission-Critical Resupply

The first issue to consider in this mission area is how soon is "soon enough." For this analysis, we envisioned a very responsive delivery system. We assume that each battalion is visited by a TS/MC resupply aircraft three times a day, at 8-hour intervals. The TS/MC resupply aircraft have space and payload capacity for one pallet of supplies and space and payload for one large, heavy item, such as a 3,000-lb-plus Stryker powerpack, on each mission. Aircraft arrive every 8 hours at each of our 18 dispersed operating locations, no matter how small the payload is—even if there is no payload at all. A system of this nature ensures that the longest any unit will have to wait for a TS/MC item once it has been located and prepared to load onto a C-130 is the average delivery time plus 8 hours.

Volume and Number of Aircraft Required

Before discussing the responsiveness of our system, we first want to see how expensive it will be in terms of intratheater airlift aircraft resources.[6] Table 5.2 shows the number of aircraft required for the Afghanistan and the Indonesia vignettes in both the no-helicopter and helicopter supply cases.

The number of aircraft round trips does not depend on the type of unit being supplied or on whether or not it has an attached artillery and aviation capability because we assumed that the TS/MC resupply requirements for any unit would always be less than what a single C-47 sortie can carry. For our system, there would be 54 round trips for each type of aircraft, three round trips a day at 18 locations. In the Afghanistan vignette, 44 C-130s are required to provide responsive TS/MC resupply using standard planning factors. Relaxing this constraint could take the required number as low as 21 C-130s. For Indonesia,

---

[6] To compute the number of aircraft required to fulfill the TS/MC resupply mission, we assumed that one C-130 sortie would depart the main operating base for each battalion every 8 hours. We assumed similar parameters, as in the routine sustainment case (1 hour to load and unload, 1.5 hours maintenance time, 300 kts, 0.5 hours noncruise flying time, and an 85 percent mission capable rate). Again, similar to the routine sustainment case, we assumed a sortie rate of 5.0 (2.5 round trips) for the CH-47.

Table 5.2
Aircraft Required for Responsive Time-Sensitive, Mission-Critical Resupply

Case	Aircraft	Metric	All Forces (number)
Afghanistan vignette			
	C-130	Aircraft, planning factor	44
		Aircraft, non–planning factor	21
	CH-47	Aircraft	22
Indonesia vignette			
	C-130	Aircraft, planning factor	59
		Aircraft, non–planning factor	25
	CH-47	Aircraft	22

59 C-130s are required using standard planning factors and possibly as few as 25 without. When the Army forces were free to maneuver away from C-130 capable bases and were supplied by helicopters from these bases, we calculated that 22 CH-47s would be required to support this TS/MC resupply concept.

Given the magnitude of the intratheater airlift effort required to provide routine sustainment, the number of aircraft involved in the TS/MC resupply mission is quite modest. This suggests that, even if it is not possible to provide all supplies to ground combat forces via air delivery, providing a responsive air-centric delivery TS/MC resupply capability requires only a small fraction of the available intratheater airlift resources.

Timeliness

Timeliness is the key metric for the TS/MC resupply mission area. A number of elements contribute to the time that passes between the moment a unit requests that a key item or person be transported by the theater logistics system and the moment it arrives. Many of the elements of this time delay are related to elements of the priorities the theater commander assigns to various units and classes of supply and to the efficiency of the theater logistics command and control systems. These elements, while affecting the performance of the intratheater airlift system—especially that of the TS/MC resupply mission—are beyond the scope of this analysis. Instead of modeling these elements, we have concentrated on the air-delivery aspects of the system. Table 5.3 presents the minimum, average, and maximum delivery times we calculated for two conditions in each vignette.[7]

The delivery times in the table represent the total time required to load the requested items, fly to the L2 or L3 airfield from the L1 airfield, and unload. In the helicopter cases, more time is added to load the requested items into a CH-47D and for the CH-47D to fly to the appropriate battalion operating area. The distances are too long to fly the helicopter directly from the L1 airfield directly to the operational area. It will often be necessary to use intermediate airfields to get closer to the point of need for the supplies because fuel is not likely to be available at the battalion operational area. The helicopters will therefore need to make the round trip without refueling.

In total, these times represent the "in-transit" time for an item that arrives to be loaded just as the loading process of the appropriate aircraft starts at the L1 base. It is just as likely that the item will arrive immediately after loading has finished. In this case, the item will have to wait 8 hours to be loaded onto the next scheduled TS/MC resupply flight that departs for the appropriate unit. If calls for TS/MC resupply items arrive randomly and are uniformly distributed across time, the average time in transit will be that shown in Table 5.3 plus 4 hours, and the maximum time in transit will be the time shown plus 8 hours.

So, the modest numbers of aircraft shown in Table 5.2 can provide average in-transit times of about 7.5 to 12 hours for the two vignettes we analyzed. Maximum in-transit times range from about 11.5 to 16 hours.

---

[7] The delivery times computed for TS/MC resupply were based on the load time of the C-130, the distance to the airfield at 300 kts (again assuming an additional one-half hour of noncruise time), and an hour to unload the aircraft. When the cargo needed to be transloaded to a helicopter for final delivery, we added an additional hour to load the cargo onto the helicopter. The helicopter travel time was computed using a distance of 55 nm for the Afghanistan case and 150 nm for the Indonesia scenario at 142 kts. No helicopter unload time was used because we believed unloading the small amount of cargo we envisioned for these cases would be a fairly short process.

Table 5.3
Time-Sensitive, Mission-Critical Resupply Delivery Times

Vignette	Case	Delivery Times	
		Metric	Hours
Afghanistan	Without helicopters	Minimum	3.3
		Average	4.1
		Maximum	4.8
	With helicopters	Minimum	4.7
		Average	5.4
		Maximum	6.2
Indonesia	Without helicopters	Minimum	3.0
		Average	4.8
		Maximum	6.1
	With helicopters	Minimum	5.1
		Average	6.8
		Maximum	8.1

The total range of in-transit times is about 3.5 to 16 hours. This seems to be a reasonable level of timeliness especially when we consider that, for a system using twice as many aircraft making flights to every unit once every 4 hours, average in-transit times would be between 5.5 and 10 hours. In this case, a 100-percent increase in resources devoted to this mission reduces in-transit time by only 21 percent. Further, this number for aircraft devoted to this mission is probably on the conservative side (i.e., could likely be reduced while providing the same level of responsiveness). Innovative CONOPS, such as committing a small number of aircraft to ground or airborne alert or holding pallet positions open on scheduled deliveries until just before flight time for TS/MC resupply mission, could allow similar responsiveness with fewer aircraft.

CHAPTER SIX
Conclusions

Several recent detailed studies commissioned by the Department of Defense, including the MCS, have established a minimum C-130 fleet size of 395. Currently, the USAF has about 490 C-130s. However, a large and growing portion of the fleet is facing flight restrictions or grounding because of fatigue-related cracking of CWB structures. If current restriction and grounding policies remain in effect, the available C-130 fleet will, within the next decade or so, no longer be able to provide the minimum intratheater airlift capability established by the MCS. This suggested performing an FSA to determine how to address this looming capability shortfall.

Several other factors could increase the needed capability beyond what the recent intratheater lift studies have established. These factors include the desire to minimize vulnerable ground movements in counterinsurgency environments, the dispersed nature of global war on terror operations, and emerging Army CONOPs that stress operational maneuver and resupply by air.

Although large multi-BCT forces operating without a ground LOC are not the current Army concept for future operations, the trend is toward more-dispersed ground-force operations. Our analysis found that routine supply of a ground combat force of moderate size using the existing intratheater airlift system is challenging. In most of the cases we analyzed, the number of C-130s required to supply six BCTs by air was equal to or greater than that likely to be available to support any one operation. Adoption of routine resupply of a multi-BCT Army unit for an extended period as an intratheater airlift task would require additional airlift assets.

The CH-47D helicopter fleet faces even greater challenges. In the case of Indonesia, CH-47 range limitations and sparse airfield availability combine to mean that the overall intratheater airlift fleet could support Army combat units in only about 55 percent of the country.

The existing intratheater airlift system can provide robust TS/MC resupply of a sizable ground force with a relatively small commitment of airlift assets. However, allocating more resources to this mission than the levels we chose offers rapidly diminishing returns: Doing so does not further reduce in-transit times. This, combined with the fact that time in transit accounts for only part of the total time between request and delivery, suggests that it may be more fruitful to invest in improving logistics management processes and procedures to substantially improve overall TS/MC resupply performance.

Because routine resupply is not a requirement and because TS/MC takes relatively few assets, the FSA should focus on ensuring that the intratheater airlift fleet continues to meet the requirement for 395 C-130s identified in the MCS. This requirement needs to be met despite the large number of aircraft that are expected to undergo flight restrictions and groundings during the next two decades.

APPENDIX

Methodology for Determining Required Daily Sustainment

This appendix describes the methodology we used to determine the required daily sustainment for the Army units we considered in this analysis. The overall methodology was based on U.S. Army CASCOM operational logistics planning data.[1] The requirement was determined for the brigade-size units in the "attack" as defined by CASCOM. We analyzed three unit types: a Stryker BCT, an infantry BCT, and a heavy BCT. We considered an augmented unit for each of these units that included additional fire support and an aviation element. The CASCOM planning factors determined the sustainment requirement for each class of supplies.

Table A.1 presents the different classes of supplies, and Table A.2 lists the unit equipment for each brigade. Tables A.3 through A.5 present the additional equipment for the augmented versions of the Stryker, infantry, and heavy units. These equipment sets are based on real-world units.

[1] The sustainment data presented in this appendix were based on the CASCOM methodology developed by an FY05 RAND Arroyo center project, unpublished RAND research.

Table A.1
Classes of Supplies

Class	Supplies
I	Subsistence, gratuitous health and comfort items.
II	Clothing, individual equipment, tentage, organizational tool sets and kits, hand tools, unclassified maps, administrative and housekeeping supplies and equipment.
III	Petroleum, fuels, lubricants, hydraulic and insulating oils, preservatives, liquids and gases, bulk chemical products, coolants, deicer and antifreeze compounds, components and additives of petroleum and chemical products, and coal.
IV	Construction materials, including installed equipment, and all fortification and barrier materials.
V	Ammunition of all types, bombs, explosives, mines, fuzes, detonators, pyrotechnics, missiles, rockets, propellants, and associated items.
VI	Personal demand items (such as health and hygiene products, soaps and toothpaste, writing material, snack food, beverages, cigarettes, batteries, and cameras—nonmlitary sales items).
VII	Major end items such as launchers, tanks, mobile machine shops, and vehicles.
VIII	Medical materiel including repair parts peculiar to medical equipment.
IX	Repair parts and components to include kits, assemblies, and subassemblies (repairable or non-repairable) required for maintenance support of all equipment.
X	Material to support nonmlitary programs such as agriculture and economic development (not included in classes I through IX).

SOURCE: Adapted from Table 6.1, U.S. Army Field Manual 4-0 (FM 10-100), *Combat Service Support*, Headquarters, Department of the Army, Washington, D.C., August 29, 2003. It is identical to our Table 4.1.

Table A.2
Unit Equipment: Stryker, Infantry, and Heavy Brigade

Stryker Brigade		Infantry Brigade		Heavy Brigade	
Stryker FOV	302	DEUCE	2	Abrams MBT	44
HMMWV	423	Forklifts	7	Bradley FOV	72
MTV	160	HEMMTT	60	M113 FOV	125
HEMTT	109	HMMWV	612	M109 FOV (M109A6 × 16)	32
Tractor	12	LMTV	124	M88	25
Forklift	8	MTV	114	ACE	6
Fox NBC	3	Bucketloaders	4	HMMWV	496
Generators (ERC A)	57	Generators (ERC A)	92	LMTV	129
Howitzer 155T	12	Howitzer light towed: M119	16	HEMMTT	120
Mortar 60 mm	18	Mortar 60 mm	14	MTV	104
Mortar 81 mm	12	Mortar 81 mm	8	Forklifts	7
Mortar 120 mm	36	Mortar 120 mm	12	5T	5
Machine gun grenade 40 mm: MK19 MOD III	118	Javelin	76	Bucketloaders	4
Machine gun: 7.62 mm	184	Machine gun grenade 40 mm: MK19 MOD III	69	Fox NBC	2
Machine gun caliber .50	306	Machine gun: 7.62 mm M240B	139	Generators (ERC A)	87
Machine gun 5.56 mm M249	279	Machine gun caliber .50	16	Rifle 5.56 mm	3,102
Rifle 5.56 mm: M4/M16A2	3,346	Machine gun 5.56 mm	350	Machine gun 5.56 mm	372
Shotgun 12 gauge	81	Rifle 5.56 mm: M4	3,041	Machine gun: 7.62 mm	287

Table A.2—Continued

Stryker Brigade		Infantry Brigade		Heavy Brigade	
Personnel	3,872	Shotgun 12 gauge	178	Machine gun caliber .50	235
Division personnel	885	Personnel	3,431	Grenade 40 mm: MK19 MOD III	71
		Division personnel	636	Mortar 120 mm	12
				Personnel	3,787
				Division personnel	885

Table A.3
Additional Equipment Required for Augmented Stryker Brigade

Unit	Amount
HIMARS battalion UA	
Equipment	
HMMWV	58
LMTV	17
MTV (18 HIMARS)	66
Generators (ERC A)	0
Machine gun caliber .50: HB flexible (ground and vehicle) W/E	9
Machine gun grenade 40 mm: MK19 MOD III	10
Machine gun: light 5.56 mm M249	46
Pistol 9 mm automatic: M9	15
Rifle 5.56 mm	376
Personnel	393
155 T BN	
Equipment	
HMMWV	61
LMTV	16
MTV	40
Generators (ERC A)	0

Table A.3—Continued

Unit	Amount
Howitzer medium towed: 155 mm M198	18
Machine gun caliber .50: HB flexible (ground and vehicle) W/E	5
Machine gun grenade 40 mm: MK19 MOD III	4
Machine gun: light 5.56 mm M249	36
Pistol 9 mm automatic: M9	15
Rifle 5.56 mm	448
Personnel	464
HQ fires UA equipment	
Equipment	
HMMWV	25
LMTV	5
MTV (18 HIMARS)	4
Generators (ERC A)	4
Machine gun caliber .50: HB flexible (ground and vehicle) W/E	1
Machine gun grenade 40 mm: MK19 MOD III	1
Machine gun: light 5.56 mm M249	8
Pistol 9 mm automatic: M9	13
Rifle 5.56 mm	93
Personnel	112
TAB (2 x Q-37)	
Equipment	
HMMWV	12
LMTV	1
MTV	4
Generators (ERC A)	4
Machine gun caliber .50: HB flexible (ground and vehicle) W/E	0
Machine gun grenade 40 mm: MK19 MOD III	1
Machine gun: light 5.56 mm M249	9
Pistol 9 mm automatic: M9	1
Rifle 5.56 mm	47
Personnel	48

Table A.3—Continued

Unit	Amount
3 ID aviation UA	
Equipment	
AH-64	48
UH-60	50
CH-47	12
HEMMTT (60 tankers)	118
HMMWV	348
LMTV	93
MTV	123
Generators (ERC A)	55
Launcher guided missile aircraft XM299: (HELLFIRE)	192
Launcher rocket aircraft: 2.75 inch 19-tube M261	96
Machine gun 7.62 mm: aircraft light door MTD	112
Machine gun caliber .50: HB flexible (ground and vehicle) W/E	90
Machine gun grenade 40 mm: MK19 MOD III	42
Machine gun: light 5.56 mm M249	214
Pistol 9 mm automatic: M9	611
Rifle 5.56 mm	2,213
Personnel	2,682

Table A.4
Additional Equipment Required for Augmented Infantry Brigade

Unit	Amount
155 T BN	
Equipment	
HMMWV	61
LMTV	16
MTV	40
Generators (ERC A)	0
Howitzer medium towed: 155 mm M198	18
Machine gun caliber .50: HB flexible (ground and vehicle) W/E	5
Machine gun grenade 40 mm: MK19 MOD III	4
Machine gun: light 5.56 mm M249	36

Table A.4—Continued

Unit	Amount
Pistol 9 mm automatic: M9	15
Rifle 5.56 mm	448
Personnel	464
HQ fires UA	
Equipment	
HMMWV	25
LMTV	5
MTV (18 HIMARS)	4
Generators (ERC A)	4
Machine gun caliber .50: HB flexible (ground and vehicle) W/E	1
Machine gun grenade 40 mm: MK19 MOD III	1
Machine gun: light 5.56 mm M249	8
Pistol 9 mm automatic: M9	13
Rifle 5.56 mm	93
Personnel	112
TAB (2 x Q-37)	
Equipment	
HMMWV	12
LMTV	1
MTV	4
Generators (ERC A)	4
Machine gun caliber .50: HB flexible (ground and vehicle) W/E	0
Machine gun grenade 40 mm: MK19 MOD III	1
Machine gun: light 5.56 mm M249	9
Pistol 9 mm automatic: M9	1
Rifle 5.56 mm	47
Personnel	48
10 MTN aviation UA	
Equipment	
OH-58	60
UH-60	50
CH-47	12
HEMMTT (60 tankers)	132

Table A.4—Continued

Unit	Amount
HMMWV	347
LMTV	8
MTV	212
Generators (ERC A)	89
Machine gun 12.7 mm: 3 barrel gatling gun XM322	54
Launcher guided missile aircraft XM292: (ATAS)	30
Launcher rocket aircraft: 2.75 inch 7-tube M260	120
Launcher guided missile aircraft	54
Machine gun 7.62 mm: aircraft light door MTD	112
Machine gun caliber .50: HB flexible (ground and vehicle) W/E	88
Machine gun grenade 40 mm: MK19 MOD III	41
Machine gun: light 5.56 mm M249	207
Pistol 9 mm automatic: M9	628
Rifle 5.56 mm	2,008
Personnel	3,022

Table A.5
Additional Equipment Required for Augmented Heavy Brigade

Unit	Amount
155 SP BN GS/R	
Equipment	
M109A6 SP	18
CATV (109 FOV)	18
CP armor carrier (M113 FOV)	9
Recovery VEH (tracked) M88	4
HEMTT (3 tankers)	5
HMMWV	52
LMTV	20
MTV	0
PLS	18
Generators (no ERC A)	
Howitzer SP 155M	18
Machine gun caliber .50: HB flexible (ground and vehicle) W/E	28

Table A.5—Continued

Unit	Amount
Machine gun grenade 40 mm: MK19 MOD III	23
Machine gun: light 5.56 mm M249	30
Pistol 9 mm automatic: M9	25
Rifle 5.56 mm	468
Personnel	486
MLRS battalion UA	
Equipment	
Launcher rocket: armored vehicle–mounted	18
CP armor carrier (M113 FOV)	12
Recovery VEH (tracked) M88	4
HEMTT	46
HMMWV	11
LMTV	18
MTV	3
PLS	6
Generators (no ERC A)	0
Machine gun caliber .50: HB flexible (ground and vehicle) W/E	15
Machine gun grenade 40 mm: MK19 MOD III	10
Machine gun: light 5.56MM M249	46
Pistol 9 mm automatic: M9	23
Rifle 5.56 mm	383
Personnel	399
75 HQ fires UA	
Equipment	
HMMWV	26
LMTV	0
MTV	4
Generators (ERC A only)	4
Machine gun caliber .50: HB flexible (ground and vehicle) W/E	1
Machine gun grenade 40 mm: MK19 MOD III	0
Machine gun: light 5.56 mm M249	8
Pistol 9 mm automatic: M9	12
Rifle 5.56 mm	108

Table A.5—Continued

Unit	Amount
Personnel	120
TAB (2 x Q-37)	
Equipment	
HMMWV	12
LMTV	1
MTV	4
Generators (ERC A)	4
Machine gun caliber .50: HB flexible (ground and vehicle) W/E	0
Machine gun grenade 40 mm: MK19 MOD III	1
Machine gun: light 5.56 mm M249	9
Pistol 9 mm automatic: M9	1
Rifle 5.56 mm	47
Personnel	48
3 ID aviation UA	
Equipment	
AH-64	48
UH-60	50
CH-47	12
HEMMTT (60 tankers)	118
HMMWV	348
LMTV	93
MTV	123
Generators (ERC A)	55
Launcher guided missile aircraft XM299: (HELLFIRE)	192
Launcher rocket aircraft: 2.75 inch 19-tube M261	96
Machine gun 7.62 mm: aircraft light door MTD	112
Machine gun caliber .50: HB flexible (ground and vehicle) W/E	90
Machine gun grenade 40 mm: MK19 MOD III	42
Machine gun: light 5.56 mm M249	214
Pistol 9 mm automatic: M9	611
Rifle 5.56 mm	2,213
Personnel	2,682

Table A.6 presents the number of tons and 463L pallets for each class of supply for the three units considered. The data are totaled in terms of dry and liquid cargo because aircraft carrying palletized cargo often reach their volume limits before they reach their payload limits; liquids are much more dense, so the converse is typically true. Table A.7 presents the required daily sustainment for the augmented units.

During this CBA, we were struck by the small amount of class V cargo (ammunition) resulting from the CASCOM analysis. In some cases, the entire daily amount of ammunition could be expended in a few minutes of intense combat. We were concerned that, in some cases, the class V daily requirement could exceed the CASCOM planning factor estimate. Since the objective of this FNA was to determine whether a capabilities gap existed, we wanted a more-conservative (larger) estimate of class V sustainment for our analysis. This would represent days with relatively high ammunition expenditure in response to high levels of hostility. We used a quarter of the "ammunition basic load" in place of the CASCOM class V planning factor estimate throughout our analysis. Table A.8 presents our estimate of the basic load for each equipment type in this analysis. We computed tons of ammunition for each unit type based on the equipment in each unit. Tables A.9 and A.10 present the daily sustainment for each unit and each augmented unit, respectively.

Finally, we assumed that each brigade consisted of three battalions and a headquarters element. We assumed that each battalion would require 30 percent of the daily sustainment of a brigade and that the headquarters would require 10 percent of the brigade sustainment.

Table A.6
Unit Daily Sustainment Using U.S. Army Combined Arms Support Command Planning Factors

	Stryker Brigade		Infantry Brigade		Heavy Brigade	
Cargo Class	463L Pallets	Tons	463L Pallets	Tons	463L Pallets	Tons
Bulk						
I	5.25	10.79	5.25	9.60	5.00	10.29
II	3.75	3.00	3.25	2.60	3.75	2.80
III (pkg)	1.00	0.90	1.25	0.90	3.25	3.60
IV	16.00	16.60	14.25	14.70	15.25	15.80
V	3.24	12.38	1.75	6.69	5.40	20.64
VI	3.25	3.00	2.75	2.60	3.00	2.80
VII	4.50	4.90	6.00	5.50	14.75	20.20
VIII	0.75	0.30	0.75	0.30	0.75	0.30
IX	12.00	12.30	2.75	2.50	13.25	13.30
Total	49.74	64.17	38.00	45.39	64.40	89.73
Liquid	**Gallons**	**Tons**	**Gallons**	**Tons**	**Gallons**	**Tons**
III (bulk)	28,183	94.41	23,228	77.81	97,885	327.91
Water	19,087	76.35	11,536	46.14	19,087	76.35
Total	47,271	170.76	34,764	123.96	116,972	404.26

Table A.7
Daily Sustainment for Augmented Units (Unit Plus Artillery and Aviation) Using U.S. Army Combined Arms Support Command Planning Factors

	Stryker Brigade		Infantry Brigade		Heavy Brigade	
Cargo Class	463L Pallets	Tons	463L Pallets	Tons	463L Pallets	Tons
Bulk						
I	9.50	19.11	9.50	17.92	9.00	15.84
II	6.75	5.20	6.25	4.80	6.50	4.90
III (pkg)	2.25	1.90	2.50	1.90	4.75	6.40
IV	28.25	29.50	26.50	27.60	26.50	27.40
V	6.88	26.29	5.39	20.60	11.14	42.59
VI	5.75	5.20	5.25	4.80	5.50	4.90
VII	23.00	13.90	24.50	14.50	34.25	32.20
VIII	1.50	0.60	1.50	0.60	1.50	0.50
IX	15.50	15.40	6.25	5.60	17.50	17.80
Total	99.38	117.10	87.64	98.32	116.64	152.53
Liquid	**Gallons**	**Tons**	**Gallons**	**Tons**	**Gallons**	**Tons**
III (bulk)	71,977	241.12	67,021	224.52	184,505	618.09
Water	29,093	116.37	21,542	86.17	28,139	112.56
Total	101,069	357.49	88,563	310.69	212,644	730.65

Table A.8
Basic Ammunition Load

Equipment Type	Basic Load (lb)	Brigades			Augmented Brigades		
		Stryker	Infantry	Heavy	Stryker	Infantry	Heavy
M-1	2,000			44			44
M-2	1,500			72			72
155 T	17,375	12			30	18	
M-109	17,375			32			50
Howitzer 155SP	17,375						18
M119A1	8,547		16			16	
Mortars							
120 mm	4,646	36	12	12	36	12	12
81 mm	1,109	12	8		12	8	
60 mm	385	18	14		18	14	
M113	600			125			125
Strykers	900	302			302		
HIMARS	36,558				18		
MLRS	64,992						18
AH-64	8,240				48		48
OH-58	2,000					60	
Troops	10	4,757	4,067	4,672	8,456	7,713	8,407
Total (tons)		358	124	465	1,059	358	1,579

Table A.9
Baseline Daily Sustainment for Each Unit

	Stryker Brigade		Infantry Brigade		Heavy Brigade	
Cargo Class	463L Pallets	Tons	463L Pallets	Tons	463L Pallets	Tons
Bulk						
I	5.25	10.79	5.25	9.60	5.00	10.29
II	3.75	3.00	3.25	2.60	3.75	2.80
III (pkg)	1.00	0.90	1.25	0.90	3.25	3.60
IV	16.00	16.60	14.25	14.70	15.25	15.80
V	23.39	89.42	8.09	30.93	30.39	116.18
VI	3.25	3.00	2.75	2.60	3.00	2.80
VII	4.50	4.90	6.00	5.50	14.75	20.20
VIII	0.75	0.30	0.75	0.30	0.75	0.30
IX	12.00	12.30	2.75	2.50	13.25	13.30
Total	69. 89	141.21	44.34	69.63	89.39	185.27
Liquid	**Gallons**	**Tons**	**Gallons**	**Tons**	**Gallons**	**Tons**
III (bulk)	28,183	94.41	23,228	77.81	97,885	327.91
Water	19,087	76.35	11,536	46.14	19,087	76.35
Total	47,271	170.76	34,764	123.96	116,972	404.26

NOTE: The difference between this table and Table A.6 is the class V requirement.

Table A.10
Baseline Daily Sustainment for Each Augmented Unit

	Stryker Brigade		Infantry Brigade		Heavy Brigade	
Cargo Class	463L Pallets	Tons	463L Pallets	Tons	463L Pallets	Tons
Bulk						
I	9.50	19.11	9.50	17.92	9.00	15.84
II	6.75	5.20	6.25	4.80	6.50	4.90
III (pkg)	2.25	1.90	2.50	1.90	4.75	6.40
IV	28.25	29.50	26.50	27.60	26.50	27.40
V	69.28	264.84	23.43	89.58	103.25	394.71
VI	5.75	5.20	5.25	4.80	5.50	4.90
VII	23.00	13.90	24.50	14.50	34.25	32.20
VIII	1.50	0.60	1.50	0.60	1.50	0.50
IX	15.50	15.40	6.25	5.60	17.50	17.80
Total	161.78	355.64	105.68	167.30	208.75	504.65
Liquid	**Gallons**	**Tons**	**Gallons**	**Tons**	**Gallons**	**Tons**
III (bulk)	71,977	241.12	67,021	224.52	184,505	618.09
Water	29,093	116.37	21,542	86.17	28,139	112.56
Total	101,069	357.49	88,563	310.69	212,644	730.65

NOTE: The difference between this table and Table A.7 is the class V requirement.

Bibliography

Air Mobility Command, U.S. Air Force, *Air Mobility Master Plan*, AMC/A55PL, October 2004.

Air Force Pamphlet 10-1403, "Air Mobility Planning Factors," undated.

Amouzegar, Mahyar A., Robert S. Tripp, Ronald G. McGarvey, Edward W. Chan, and C. Robert Roll, Jr., *Supporting Air and Space Expeditionary Forces: Analysis of Combat Support Basing Options*, Santa Monica, Calif.: RAND Corporation, MG-261-AF, 2004. As of June 23, 2009:
http://www.rand.org/pubs/monographs/MG261.html

Bush, George W., *The National Security Strategy of the United States of America*, Washington, D.C.: The White House, September 2002.

Chairman of the Joint Chiefs of Staff, Instruction (CJCSI) 3170.01E, *Joint Capabilities Integration and Development System*, May 11, 2005.

———, CJCSI 3170.01F, *Joint Capabilities Integration and Development System*, May 1, 2007.

———, CJCSI 3170.01G, *Joint Capabilities Integration and Development System*, March 1, 2009.

———, Manual (CJCSM) 3170.01C, *Operation of the Joint Capabilities Integration and Development System*, May 1, 2007.

CJCSI—*See* Chairman of the Joint Chiefs of Staff, Instruction.

Department of Defense Instruction (DoDI) 5000.2, *Operation of the Defense Acquisition System*, May 12, 2003.

———, *Operation of the Defense Acquisition System*, December 8, 2008.

GlobalSecurity, "CH-47D Chinook," Web page, September 22, 2005. As of June 23, 2009:
http://www.globalsecurity.org/military/systems/aircraft/ch-47d.htm

Headquarters Department of the Army, *Technical Manual for Army CH-47D Helicopter*, January 31, 2003.

Headquarters U.S. Air Force, Future Concepts and Transformation Division, *The U.S. Air Force Transformation Flight Plan 2004*, Washington, D.C., January 1, 2004.

Joint Chiefs of Staff, *The National Security Strategy of the United States of America*, September 2002.

———, *Focused Logistics Campaign Plan*, 2004.

———, *Joint Logistics (Distribution) Joint Integrating Concept*, Vers. 1.0, February 7, 2006.

Joint Chiefs of Staff, Joint Experimentation, Transformation, and Concepts Division (JS/J7), *Capstone Concept for Joint Operations*, Vers. 2.0, August 2005.

———, *JOpsC Family of Joint Concepts—Executive Summaries*, briefing, August 23, 2005.

Joint Chiefs of Staff, Force Application Assessment Division (JS/J8), *Joint Capabilities Integration and Development System (JCIDS)*, January 2006.

Orletsky, David T., Daniel M. Norton, Anthony D. Rosello, William Stanley, Michael Kennedy, Michael Boito, Brian G. Chow, and Yool Kim, *Intratheater Airlift Functional Solution Analysis (FSA)*, Santa Monica, Calif.: RAND Corporation, MG-818-AF, 2011. As of February 3, 2011: http://www.rand.org/pubs/monographs/MG818.html

Orletsky, David T., Anthony D. Rosello, and John Stillion, *Intratheater Airlift Functional Area Analysis (FAA)*, Santa Monica, Calif.: RAND Corporation, MG-685-AF, 2011. As of February 3, 2011: http://www.rand.org/pubs/monographs/MG685.html

Training and Doctrine Command (TRADOC) Pamphlet 525-3-1, "United States Army Operational Concept for Operational Maneuver 2015–2024," Vers. 1.0, Fort Monroe, Va.: Headquarters, U.S. Army Training and Doctrine Command, October 2, 2006.

U.S. Air Force, "Global Mobility CONOPS," Vers. 4.3, working draft, December 29, 2005.

U.S. Army Field Manual 4-0 (FM 10-100), *Combat Service Support*, Headquarters, Department of the Army, Washington, D.C., August 29, 2003.

U.S. Army, Training and Doctrine Command Analysis Center, *Future Cargo Aircraft (FCA) Analysis of Alternatives (AoA)*, July 18, 2005, Not Available to the General Public.

U.S. Army and U.S. Air Force, "Way Ahead for Convergence of Complementary Capabilities," memorandum of understanding, February 2006.

U.S. Army Aviation Center, Futures Development Division, Directorate of Combat Developments, *Army Fixed Wing Aviation Functional Area Analysis Report*, Fort Rucker, Ala., June 3, 2003a.

———, *Army Fixed Wing Aviation Functional Needs Analysis Report*, Fort Rucker, Ala., June 23, 2003b.

———, *Army Fixed Wing Aviation Functional Solution Analysis Report*, Fort Rucker, Ala., June 8, 2004.

U.S. Department of Defense, *Joint Operations Concepts*, November 2003a.

———, *Focused Logistics Joint Functional Concept*, Vers. 1.0, Washington, D.C., December 2003b.

———, *Homeland Security Joint Operating Concept*, Washington, D.C., February 2004a.

———, *Strategic Deterrence Joint Operating Concept*, Washington, D.C., February 2004b.

———, *Force Application Functional Concept*, Washington, D.C., March 5, 2004c.

———, *Stability Operations Joint Operating Concept*, Washington, D.C., September 2004d.

———, *Major Combat Operations Joint Operating Concept*, Washington, D.C., September 2004e.

———, *Joint Forcible Entry Operations Joint Integrating Concept*, Vers. 92A3, limited distribution draft, Washington, D.C., September 15, 2004f.

———, *The National Defense Strategy of the United States of America*, Washington, D.C., March 2005a.

———, *Seabasing Joint Integrating Concept*, Vers. 1.0, Washington, D.C., August 1, 2005b.

———, *Homeland Defense and Civil Support Joint Operating Concept*, Vers. 1.5 (draft), November 2005c.

———, *Quadrennial Defense Review Report*, Washington, D.C., February 6, 2006.

———, *The National Defense Strategy of the United States of America*, Washington, D.C., June 2008.

U.S. Department of Defense and the Joint Chiefs of Staff, *Mobility Capabilities Study*, Washington, D.C., December 2005, Not Available to the General Public.

Vick, Alan, David Orletsky, Bruce Pirnie, and Seth Jones, *The Stryker Brigade Combat Team: Rethinking Strategic Responsiveness and Assessing Deployment Options*, Santa Monica, Calif.: RAND Corporation, MR-1606-AF, 2002. As of June 23, 2009:
http://www.rand.org/pubs/monograph_reports/MR1606.html

Warner Robins Air Logistics Center, ALC/LB, "C-130 Center Wing Status," briefing, February 9, 2005.

Yarnell, Edward, "Joint Transformation Concepts," briefing, Joint Experimentation, Transformation and Concepts Division (J7), January 6, 2006.

CPSIA information can be obtained at www.ICGtesting.com
Printed in the USA
LVOW102034070113

314714LV00032B/1886/P

9 780833 047557